A First Guide to Statistical Computations in R

Torben Martinussen
Ib M. Skovgaard
Helle Sørensen

Torben Martinussen, Ib Michael Skovgaard and Helle Sørensen

A First Guide to Statistical Computations in R

1. edition 2012, Biofolia

© Biofolia

Cover: Sofie Meedom
Typeset: Torben Martinussen, Ib Michael Skovgaard and Helle Sørensen
Print: Narayana Press, Gylling, Denmark

ISBN 978-87-913-1956-3

Published by:
Samfundslitteratur
Rosenoerns Allé 9 DK-1970 Frederiksberg C
Denmark
Tlf: + 45 38 15 38 80
Fax: + 45 35 35 78 22
slforlagene@samfundslitteratur.dk
www.biofolia.dk

All rights reserved.
No parts of this book may be reproduced or transmitted in any form or by any means, electronic or mechanical, including photocopying, recording, or by any information storage or retrieval system, without permission in writing from the publisher.

Preface

R is a statistical computer program used and developed by statisticians around the world. It is probably the leading statistical program, at least among statisticians, and it is freely available [R Development Core Team, 2010]. You can read more about the program on the R home page www.r-project.org.

This guide to R is intended for the newcomer who wants to do statistical analysis with R, and needs a guide to get started and a reference for common data handling and statistical analysis. The guide is in no way complete, but you should be able to do quite a bit based on this guide alone. Once acquainted with R you will surely need information from the numerous other sources, many of which are freely available and easily found from the R pages. Notice that R can also be used as a general purpose programming language, but this is outside the scope of this guide.

The selection of material reflects what is needed for our teaching in statistics for students of various disciplines, mostly biological, at the University of Copenhagen.

The guide is divided into two parts, the first part on R basics, and the second part on statistical analyses using R. In the second part we use various datasets for illustration. They are all available in the R package Guide1data, see Appendix C for installation of the package. There is also a supporting web site

http://www.r-guide.dk

where you among other things find most of the R-code used in the book.

To get started you should not read from one end to the other, but first install R (see Appendix A), then read the first few sections in the first chapter while making sure that you can redo what is shown, and as soon as possible start working on your own data, reading sections as needed. We have tried to make the purpose of each chapter and section clear so that it is easier for you to select what to read. This is done explicitly in Part I with an introductory remark to each section or subsection (in italic), whereas it is more obvious in Part II.

The way you work with R does not depend much on the operating system, and we have chosen the few such instances to describe how it works in Windows in the hope that the analogue can be guessed by users of other platforms.

To improve the layout, the graphs in the book have been produced with changed graphical parameters, such as the size of plot symbols and labels, rather than with the default values. The exception to this is Section 5.2.1 where some of these options are described. Some graphs use colors, but as the book has been typeset in greyscale the colors are not shown in the book. The colored versions of these graphs can be found at the supporting web site. Moreover, the output from R has been edited a few places to make it more readable.

Contents

I	**R basics**	**9**
1	**Working with R**	**11**
	1.1 The script-editor	11
	1.2 R as a pocket calculator	13
	1.3 A small example program: Potato yield	13
	1.4 Variables and variable names	16
	1.5 Functions in R	16
2	**Getting your data into R**	**19**
	2.1 Copy the data via the clipboard	19
	2.2 Save and read data in the csv format	20
	2.3 Read an Excel sheet directly	20
	2.4 Make R find the file	21
	2.5 Data frames	21
3	**Types and structures of data in R**	**23**
	3.1 Numeric values and character values	23
	3.2 Vectors in R	24
	3.3 Logical values	27
	3.4 Missing values	28
	3.5 Matrices	29
4	**Data handling**	**33**
	4.1 How to select parts of a data frame	33
	4.2 Combining data	34
	4.3 Analysis of data split into subgroups	35
	4.4 Maintaining order	36

4.5 How to work on a data frame 37

II R for statistical analysis 39

5 R functions for data presentation 41
5.1 Functions listing data or data summaries 41
5.2 Graphical functions 43

6 Basic statistical functions and tests 49
6.1 The normal distribution 49
6.2 Parametric two-sample tests 51
6.3 Discrete distributions 54
6.4 Tests in contingency tables 57
6.5 Non-parametric tests 62
6.6 Power and sample size calculations 65

7 Linear normal models 69
7.1 Analysis of variance (ANOVA) 69
7.2 Regression-type models 78
7.3 Model validation 83
7.4 Estimation of contrasts 86

8 Models with random effects 91
8.1 F-tests and likelihood ratio tests 91
8.2 Analysis of linear mixed models (`lme`, `lmer`) 93

9 Repeated measurements 103
9.1 Preliminaries 103
9.2 Analysis of summary measures 105
9.3 The random intercepts model 106
9.4 Investigation of the correlation structure 107
9.5 Serial correlation and variance homogeneity 110
9.6 Models with variance inhomogeneity 118
9.7 Multiple series for each subject 124

10 Generalized linear models 129
10.1 Logistic regression 129
10.2 Proportional odds model 137
10.3 Poisson regression 139

11 Non-linear regression 145
11.1 Non-linear regression with `nls` 145

11.2 Model validation, transform-both-sides 149
11.3 Estimation of derived parameters 152

12 Survival analysis   **153**
12.1 Survival data 153
12.2 Kaplan-Meier estimator, log-rank test 154
12.3 The Cox proportional hazards model 156
12.4 Time-varying effects 158

A How to install R   **161**

B Add-on packages   **163**

C The data package `Guide1data`   **165**

D Getting help and further information   **167**

Bibliography   **169**

Part I

R basics

This part introduces you to the R system and gives you the basics for using R. We recommend that you read through the sections you need and return to the other ones when need arises. Notice how each section starts with a short description of the content. First of all, before even reading this part, you should download and install R on your computer, see Appendix A.

1
Working with R

We assume that R has already been installed, if not, see Appendix A for installation of R. The following description applies to Windows, but most is similar for other systems.

You then start R by double-clicking on the R icon (a capital R) on the desktop.

The R console window, see Figure 1.1, then appears and you meet the prompt

    >

which means that R is ready to receive a command. For example, you write 3+2

    > 3+2
    [1] 5

and R responds with the second line when you press Enter. You quit R via the File menu or with the command q().

1.1 The script-editor

Read this when you want to know how to be able to save your work, load it again, make corrections, and so on.

In practice when you are using R on your own data, you will want to save the commands, so that you can repeat what you have done. Therefore it is recommended that you write your commands in a so-called R script (or R program) and then run the commands from the script. To do that choose `File` followed by `New script` in the menu bar in order to open a new script, write commands in the script and run a command line by `Ctrl+R` (or click `Edit` followed by `Run line or selection`). The command is then automatically transferred to the R console, and the command as well as the result appear as before. Instead of the `Ctrl+R` command you may also use copy-paste from the R-script to the R console.

Save the R script when you finish the R session (`Ctrl+S` or via the File menu), then you have the commands for later use. You may in principle use any text-editor (Notepad, for example, but not Word as it uses its own format to save what you type) to write your commands to R and then copy-paste the text into the R console. It is most convenient to save all files (data files and R programs) regarding a certain project or course in the same directory, and ask R to use this directory as the so-called working directory. The working directory is the directory where R looks

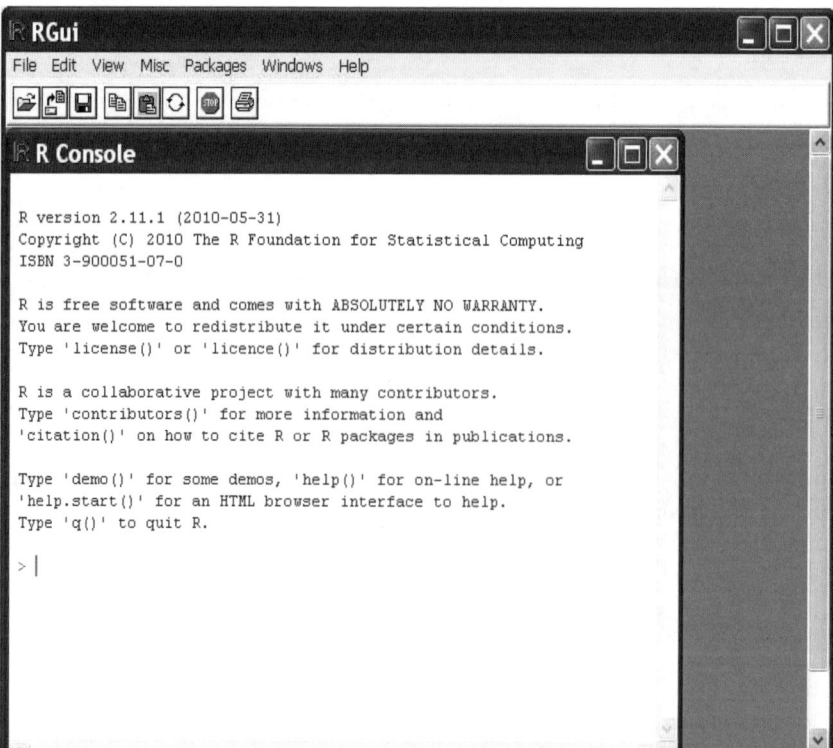

Figure 1.1. The R console window.

for files you try to read or stores files you write, unless you somehow instruct it differently. The command `getwd()` gives you the full path of the current working directory. You can change the working directory from the File menu. In any case you may save your file in any location you choose via the File menu in R by choosing File and then Save as or, if the file already exists, just File followed by Save to overwrite the old version. When you save the file you should notice in which folder you keep it so that you can find it when you next open it via File followed by Open script.

A convenient feature in a script is that you may write comments, since R will ignore everything following a "hash" sign (#) on a line. See the example program in Section 1.3.

1.2 R as a pocket calculator

This section illustrates how you may use R to do some simple calculations.

R may be used as a simple pocket calculator. All standard functions (powers, exponential, square-root, logarithm, etc.) are, of course, built-in. For example,

```
> (17-3)/(3-1) + 1
[1] 8
> 3*4 - 7 + 1.5
[1] 6.5
> 1.8^2 * exp(0.5) - sqrt(2.4)
[1] 3.792664
```

The last example shows that $1.8^2 \cdot e^{0.5} - \sqrt{2.4}$ equals 3.79.

1.3 A small example program: Potato yield

Read this if you want to see the typical process of making a small but realistic statistical analysis in R. Do not focus on details until later.

Nine plots with potatoes were fertilized in 1982 with 0, 20 or 40 units of phosphorus (P) and the yield was measured (in hkg/ha). The previous year these nine plots were fertilized with amounts 0, 30 or 60 of phosphorus. The dataset (extracted from a larger experiment) is in a text file "potatoes.txt" with the 10 lines

```
P81 P82 yield
  0   0  330
  0  20  346
  0  40  409
 30   0  368
 30  20  371
 30  40  409
 60   0  360
 60  20  382
 60  40  398
```

The following R program reads the data from the text file into R, plots yield against phosphorus amount in 1982 (P81), and fits a straight line to the points by a so-called linear regression — quite a bit in six (or actually just five) lines of code:

```
> potato = read.table(file="potatoes.txt",
+                     header=TRUE, dec=".")
> dim(potato)
> head(potato)
> plot(x=potato$P82, y=potato$yield)
> # Next a straight line is fitted to the data.
> lm(yield~P82, data=potato)
```

A brief description of what happens is given below; more details on the commands used follow later in the various contexts.

The first line uses the R function `read.table` to read the data from the file and store the result under the name `potato`. The equality sign in this context is an "assignment" operator that stores whatever is the result on the right-hand side under the name on the left.[1] The result of `read.table` is a so-called *data-frame* in R which you may think of as a number of columns with names (the headers). The "options" `header=TRUE` and `dec="."` indicate that the first line in the file contains the headers and that a period is used as decimal point. The second line makes R respond with

```
> dim(potato)
[1] 9 3
```

telling us that the data frame `potato` has 9 rows and 3 columns. Ignore

---
[1] You may use a left arrow, <-, instead of an equality sign, and experts tend to prefer that because it emphasizes that assignment has a direction, and because of slight technical differences. We use "=" for convenience.

1.3 A small example program: Potato yield

the square brackets for now. The command head(potato) causes the first few lines of the data frame potato to be printed as

```
> head(potato)
  P81 P82 yield
1   0   0   330
2   0  20   346
3   0  40   409
4  30   0   368
5  30  20   371
6  30  40   409
```

This command is very useful to see the layout of a large dataset.

Next, the command plot(x,y) creates a graph with y plotted against x where, in our case, x and y are the two "columns" with headers P82 and yield in the data frame potato. Note the use of "$" between the name of data frame and column header.

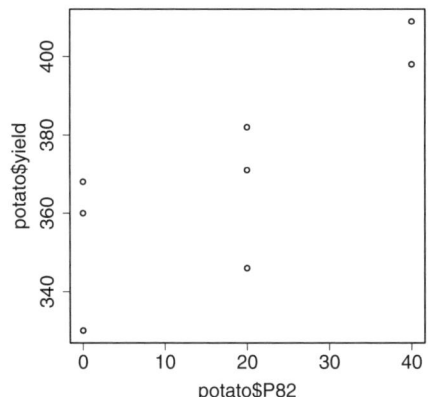

Figure 1.2. Potato yields against P-fertilization

The resulting figure appearing in a separate graph window is seen in Figure 1.2.

The fifth line is a comment ignored by R because it starts with a # and R then ignores the rest of the line. The sixth line fits a line to the points by using the R function lm. This function is used to fit statistical so-called linear models. This is a comprehensive topic that will be dealt with separately in Chapter 7. The results from this function are

```
Call:
lm(formula = yield ~ P82, data = potato)

Coefficients:
(Intercept)          P82
    348.444        1.317
```

with the fitted values of intercept and slope of the line.

1.4 Variables and variable names

Since most analyses consist of several steps it is essential to store the intermediate results. This is what variable names (or object names) are used for.

You can also assign values to named variables and use them for later computations. For example,

```
> x = 1.8^2
> y = exp(0.5)
> z = x*y - sqrt(2.4)
> z
[1] 3.792664
```

Note that when you just write z, or if you write print(z), R prints its value.

Variable names are chosen by you, and they may contain periods as in fit.data1, but not spaces. R is case sensitive, which means that y and Y will be different. Certain names are already used, for example, sum is a built-in function, so you should not use that name for your variables.

Variables need not be single numbers, but can contain an entire data frame, a vector, a matrix or the results from an analysis of data by the function lm, for example.

1.5 Functions in R

Read this when you want to know, why there is a comma or parenthesis here and not there, and to understand the logic of commands in R.

Functions perform almost anything you do in R, so it is worthwhile noting how they are used. A function has a name and one or more named

1.5 Functions in R

arguments. Thus, the function `sqrt` has a single argument which you put inside the parentheses following the name of the function: `sqrt(2.4)`, for example.

You have already seen several functions with more than one argument, for example

```
> read.table(file="potatoes.txt", header=TRUE, dec=".")
```

Here we have used three arguments: `"potatoes.txt"`, `TRUE` and `"."`.

To distinguish the roles of the different arguments, the arguments of a function have names; in the present case the three names are `file`, `header` and `dec`, while the `values` of the arguments are `"potatoes.txt"`, `TRUE` and `"."`. When you call a function in your program you write the function name followed by a left and right parenthesis with the arguments, separated by commas, in between. If you include the names of each argument, the arguments can come in any order, whereas you may omit the names if the arguments come in the right order. So we might have written

```
> read.table("potatoes.txt", header=TRUE, dec=".")
```

because the argument with the name `file` is the first in the function `read.table`. How do you know that? Well, you need to see a description of the function, which you can get by writing

```
> ?read.table
```

in the R console.

Sometimes you call a function with no arguments. Then you still need the parentheses. For example, `getwd()` gives as result the full path name of the current working directory, see Section 1.1.

For non-standard analyses you would sometimes have to write your R functions yourself. We will do so a few times in the book, but we will not go into details about the many possibilities. The book by Venables and Ripley [2000] is an excellent reference for actual programming with R.

1.5.1 Default values of arguments

Another very convenient feature with functions in R is that *default argument values* may be used. For example, in `read.table` the default value of the argument `dec` is a period, meaning that we could have omitted the part `dec= "."`, because R would then use the default value. However, be careful, because if your computer is set up to use a decimal comma, you will end up with annoying errors unless you include `dec=","` part.

2
Getting your data into R

You probably have your data in a file and need to get them into R. Here you see how.

In the introductory example you saw how the function `read.table` can be used to get read your data from a file. However, you probably do not keep your data as a plain text file, but rather as a spreadsheet such as Excel. Below you will see ways to read such data into R.

2.1 Copy the data via the clipboard

Numbers in rows and columns can just be copied, and then read in R as shown here.

Suppose, in the introductory example from Section 1.3, you had the same 10 lines of data as three columns in a sheet in Excel. As a newcomer to R you will probably appreciate an easy way to read the data into R:

- Mark the data with headers (10 lines and three columns) and copy this selection (right-click and choose "copy", or press `Ctrl-Insert`).

- In R write the command

```
> potato = read.table(file="clipboard", header=TRUE, dec=".")
```

and you have again the data frame `potato` with three variables.

2.2 Save and read data in the csv format

Comma-separated values (csv) denotes a standard format most spreadsheets can read and write. It is one recommendable way to store your data.

When you work on a project and read the same data many times you will get tired of having to open the Excel file, copy the data and so on, every time. The recommended way of avoiding that is to save your Excel data sheet as a `.csv` file, which means in the "comma-separated values" format. This was originally just a text-file with commas separating the cells, but now semicolons are often used as separator instead of commas. Suppose you have stored the file "potatoes.csv" with the data. Then, in R, you write

```
> potato = read.csv(file="potatoes.csv", sep=";", dec= ".")
```

if you have chosen semicolon as the field separator, or else you write `sep=","`. In the function `read.csv` the argument `header` has default value `TRUE`, so we can skip that argument.

2.3 Read an Excel sheet directly

Yes, you can read Excel files more directly, but you may need minor changes in your installation.

In practice you will often keep your data in an Excel spreadsheet. The are two R-packages available for reading and writing such files: `xlsReadWrite` and `xlsx`. The first only works with the 2003-version of Excel, while the second works also with the 2007-version. The use of the function `read.xlsx` from the `xlsx` package is very simple. For example, to read the first sheet in the Excel-file `potatoes.xlsx` you write

```
> potato = read.xlsx(file="potatoes.xlsx", sheetIndex=1)
```

Other useful options are `from=3`, for example, which instructs R to ignore the first two rows, `header=FALSE` if there are no column names in the file and `sheetName="Exp1"` which may be used instead of `sheetIndex` if the sheet you want to read has the name `Exp1`.

There is also a function called `write.xlsx` in the `xlsx` package.

Be prepared, however, that xlsx may not work immediately on your computer, depending on your installation and setup of Java. To see how you may resolve the problem see the help page by writing ?xlsx.

2.4 Make R find the file

Where does R look for the file, and where should you keep it?

What if R cannot find the data file potatoes.txt or potatoes.csv? You can use the command getwd() to see what is the working directory where R searches for the file, and you can change the working directory from the R menu. An alternative is to write

```
> potato = read.csv(file=file.choose(), sep=";", dec= ".")
```

which will open a window where you can browse for your data file. Note that the parentheses following file.choose are required because file.choose is a function.

2.5 Data frames

R does not have spreadsheets; it has data frames, which are in many ways similar.

A table of data read by read.table or by read.csv is called a data frame. You may think of a data frame like a spreadsheet, with rows and names columns as you have already seen.

You can get various information about a data frame using the functions dim and names

```
> potato = read.csv("potatoes.csv", sep=";", dec=".")
> dim(potato)
[1] 9 3
> names(potato)
[1] "P81" "P82" "yield"
```

giving the number of rows and columns (variables), and the names associated with the columns, respectively. Also the function head that was used in the introductory example in Section 1.3 may be useful to see the structure of the data.

To get access to the values of a column of data after it has been read into a data frame in R, you must address it "politely" by the its full name consisting of the name of the data frame and the name of the column with a dollar-sign in between, For example, to calculate the mean yield in the potato example you write

```
> mean(potato$yield)
[1] 374.7778
```

2.5.1 The functions attach, detach and with

If you want to work on a spreadsheet you open that particular one. In R you get direct access to a data frame by the function attach.

When you work on a specific dataset it is a nuisance to have to write potato$ in front of all the names. You may avoid that by writing

```
> attach(potato)
```

after which you have direct access to the short names so that you can write

```
> mean(yield)
[1] 374.7778
```

However, if you attach several data frames, or you have used some of the variable names already, name-conflicts will arise. You may close down the direct access to the names within the data frame by

```
> detach(potato)
```

but in practice the use of **attach** and **detach** cause a lot of troubles until you are more experienced. Therefore it is recommended instead to use the function **with**. For example, to plot the yield against P82 you write

```
> with( potato, plot(P82,yield) )
```

If you prefer to use **attach** it is a good idea to close down R and start it again when you start something new.

3

Types and structures of data in R

Numbers can be added, names cannot, but both types may be used in R just as in a spreadsheet´, and there are more than those two types.

Data in R may be single numbers, a column of numbers (or rather: a vector), a name or some other sequence of letters, a vector with names, and so on. A brief introduction of the most common types and shapes are given here.

3.1 Numeric values and character values

Numbers and characters are the two most important types of values.

The basic values in R are typically like the content of a single cell in a spreadsheet. This may be a number, like 31.2, but it may also be a word, like "male" or "female". The latter type is a *character* value as opposed to numbers which are numeric values. Character values are written in quotes. Otherwise, if you make an assignment

```
> varietyname = Pallas   # Wrong - should be "Pallas"
```

R will respond with an error message, because it looks for a variable by the name Pallas.

Character values may contain spaces, for example "first year" or "second year" (another reason for having the quotes around the values).

There are other types of values, first of all logical values (see below).
Make sure, perhaps after reading about vectors also, that you understand
the difference between a variable name and a character value.

3.2 Vectors in R

*Vectors are like columns in a spreadsheet — a series of values of the same
type. Here you may read how you make vectors and work with them.*

You are probably used to work with data in columns like in a spreadsheet.
In R a vector may be thought of as such a column, and the name of the
vector corresponds to your column header. Thus, in the introductory
example, `potato$yield` is a vector which you can inspect by writing its
name:

```
> potato$yield
[1] 330 346 409 368 371 409 360 382 390
```

Recall that the dollar-sign $ in `potato$yield` is used to denote the vector
(or column) `yield` in the data frame `potato`.

Vectors can be added, multiplied and so on element-wise by use of ordinary arithmetic expressions,

```
> potato$P81
[1]  0  0  0 30 30 30
> potato$P82
[1]  0 20 40  0 20 40
> potato$P81 + potato$P82
[1]  0 20 40 30 50 70
```

and built-in mathematical functions may similarly be applied to vectors.
For example,

```
> sqrt(potato$yield)
[1] 18.166 18.601 20.224 19.183 19.261 20.224 18.974
    19.545 19.748
```

is a new vector containing the nine square roots of the yields.

To read a vector into R you probably most often get it from a data frame
as shown above, but sometimes you may need to type the data. To
construct a vector that way, the function c(...) is used, and the data

3.2 Vectors in R

entries need to be separated by commas. Thus, to type in a vector with the five numbers $(2, 5, 6, 9, 17)$ you have to type[1]

```
> x = c(2,5,6,9,17)
```

and the result is seen by typing the name of the vector:

```
> x
[1]  2  5  6  9 17
```

The name c of the function stands for "concatenation", meaning that it is used to join several vectors into one single longer vector. In the example just shown the five single-element vectors 2, ..., 17 were concatenated into the vector x of length 5. The lines

```
> y = c(21, 29)
> c(x,y)
[1]  2  5  6  9 17 21 29
```

show how you concatenate the vector x of length 5 with a vector, y, of length 2.

Some other useful functions that create certain vectors are rep and seq,

```
> w = rep(2, times=6)
> w
[1] 2 2 2 2 2 2
> u = rep(c(1,2,3), each=4)
> u
 [1] 1 1 1 1 2 2 2 2 3 3 3 3
> z = seq(from=1, to=15, by=3)
> z
[1]  1  4  7 10 13
```

where seq(from=1, to=15, by=3) gives the natural numbers from 1 to 15 with an inter-distance of 3. Since the inter-distance between 13 and 15 is only 2, 15 is not included in the sequence. There is a special brief version of sequences with steps of size 1, namely

```
> y = 1:5
> y
[1] 1 2 3 4 5
```

---

[1] except if you use the function scan

Perhaps you have wondered why R starts each line with a pair of brackets, []. This is quite convenient for long vectors that fill up more than one line. For example,

```
> (1:25)^2
 [1]   1   4   9  16  25  36  49  64  81 100 121 144 169 196
[15] 225 256 289 324 361 400 441 484 529 576 625
```

shows that 400 is element number 20 in the sequence of squared integers.

3.2.1 How to select elements of a vector

One may access the entries in a vector using the [] notation. For example,

```
> attach(potato)
> yield
[1] 330 346 409 368 371 409 360 382 398
> yield[1]
[1] 330
```

gives the first element of the data vector `yield`. Several elements may be selected by using a vector of indices, for example,

```
> yield[2:4]
[1] 346 409 368
> i = c(1,4,7)
> yield[i]
[1] 330 368 360
```

A minus sign before an index-vector de-selects those elements, as in

```
> yield[-1]
[1] 346 409 368 371 409 360 382 398
> i = 7:9
> yield[-i]
[1] 330 346 409 368 371 409
```

Finally, you should notice that you may also use *logical indexing*, see the next section.

3.3 Logical values

When you need to check a condition, such as which of the observations are from a particular year, you need to know about logical values.

Certain expressions passed on to R give the answer TRUE or FALSE, for example,

```
> attach(potato)
> yield
[1] 330 346 409 368 371 409 360 382 398
> yield>400
[1] FALSE FALSE  TRUE FALSE FALSE  TRUE FALSE FALSE FALSE
```

The result of the expression yield>400 is a vector with values that are *logical*, that is, TRUE or FALSE. The first value of yield is 330 and therefore the first value of yield>400 is FALSE, and so on.

Just like numeric or character values, logical values may be stored under a name

```
> cut400 = (yield>400)
```

which in this case is a vector of length 9. This vector may then be used, for example, to extract the values of yield that fulfill the condition (that is where the vector yield>400 has the value TRUE)

```
> yield[cut400]
[1] 409 409
```

Notice that here, in the square brackets, we use a logical vector of the same length as yield as index, and that the result is a sub-vector of yield. The same result could be obtained by *direct indexing*

```
> i = c(3,6)
> i
[1]  3  6
> yield[i]
[1] 409 409
```

but the *logical indexing* in terms of the logical vector, cut400, is often more convenient. The logical vector needs not have anything to do with the vector yield as long as it has the same length. For example, still with the data from the introductory example, we could write

```
> yield[P82==0]
[1] 330 368 360
```

to select the yields of the plots without phosphorus fertilization in 1982. Notice that in such a `comparison` you must use a double equality sign.

The function `which` returns the vector with the indices of the values that are true. Thus,

```
> which(cut400)
[1] 3 6
```

informs us which cases are selected by the condition `yield>400`.

The logical operators are < (used above), <=, >, =>, ==, and != where == and != mean equal and not equal, respectively.

Generally ! is the logical `not` which reverses `TRUE` and `FALSE`, for example,

```
> !cut400
[1]  TRUE  TRUE FALSE  TRUE  TRUE FALSE  TRUE  TRUE  TRUE
```

Logical expressions may be combined using the logical operators & (and) and | (or). For example,

```
> P81
[1]  0  0  0 30 30 30 60 60 60
> P82
[1]  0 20 40  0 20 40  0 20 40
> (P81>0) & (P82>0)
[1] FALSE FALSE FALSE FALSE  TRUE  TRUE FALSE  TRUE  TRUE
>   (P81>0) | (P82>0)
[1] FALSE  TRUE  TRUE  TRUE  TRUE  TRUE  TRUE  TRUE  TRUE
> yield[(P81>0) | (P82>0)]
[1] 346 409 368 371 409 360 382 398
```

so the vector ((P81>0) & (P82>0)) has the value `TRUE` when both conditions are true while ((P81>0) | (P82>0)) has the value `TRUE` when either one or both of the two conditions are true.

3.4 Missing values

Read this short section right away, since data often are incomplete.

Often in real experiments some of the data may be missing for various reasons (experiment failed, measurement below detection limit, etc.). This is indicated in R with NA (Not Available). Suppose that the first recording of potato$yield is missing so that the vector is

```
> potato$yield
[1]  NA 346 409 368 371 409 360 382 398
```

Operations on an element with a NA-value gives the value NA as for example

```
> potato$yield[1] + potato$yield[2]
[1] NA
```

Also a function, such as sum, which normally calculates the sum of the components of the vector, returns the value NA when applied to a vector in which just one of the values is NA. However, the sum of the non-missing elements is computed if the option na.rm=TRUE is supplied to sum:

```
> sum(potato$yield)
[1] NA
> sum(potato$yield, na.rm=TRUE)
[1] 3043
```

Alternatively, you may use the function is.na:

```
> is.na(yield)
[1]  TRUE FALSE FALSE FALSE FALSE FALSE FALSE FALSE FALSE
> sum(potato$yield[!is.na(potato$yield)])
[1] 3043
```

3.5 Matrices

Unless you are programming calculations yourself, you probably don't need to know about matrices now. This is a very brief introduction.

Matrices can be constructed using for example the function matrix as shown below:

```
> A = matrix(1:6, nrow=2, ncol=3)
> B = matrix(1:6, nrow=3, ncol=2)
```

```
> A
         [,1] [,2] [,3]     # Matrix with 2 rows and 3 columns.
    [1,]   1    3    5
    [2,]   2    4    6
> B
         [,1] [,2]          # Matrix with 3 rows and 2 columns.
    [1,]   1    4
    [2,]   2    5
    [3,]   3    6
> A%*%B                     # Matrix multiplication AB
         [,1] [,2]
    [1,]  22   49
    [2,]  28   64
```

Recall that what is written on a line after a # in R is ignored and may thus be used to write comments. Notice that the matrix as default is filled column by column. You may use the byrow=TRUE as in

```
> A = matrix(1:6, nrow=2, ncol=3, byrow=TRUE)
> A
         [,1] [,2] [,3]
    [1,]   1    2    3
    [2,]   4    5    6
```

to change this to the rows. To do matrix calculations such as calculating the determinant and the inverse of matrix there are built-in functions like det(A) which calculates the determinant of a square matrix A and solve(A) which calculates the inverse matrix of A.

3.5.1 How to select elements, rows, columns of a matrix

Consider the matrix

```
> x = matrix(1:12, nrow=3, ncol=4, byrow=TRUE)
> x
         [,1] [,2] [,3] [,4]
    [1,]   1    2    3    4
    [2,]   5    6    7    8
    [3,]   9   10   11   12
```

and suppose you want to select the value in the second row and third column. With matrices, like vectors you use square brackets for direct indexing, but now with row number and column number separated by a comma,

3.5 Matrices

```
> x[2,3]
[1] 7
```

Again you may use a vector to select more rows and/or columns and a minus to de-select rows and/or columns. Some examples,

```
> x[1:2, c(2,4)]
     [,1] [,2]
[1,]    2    4
[2,]    6    8
> x[2,-3]
[1] 5 6 8
```

Also, you may leave the row (or column) specification empty to select all rows (or columns).

```
> x[2:3,]
     [,1] [,2] [,3] [,4]
[1,]    5    6    7    8
[2,]    9   10   11   12
```

Logical indexing may be used also with matrices. Just replace the index numbers in the examples above by a logical row index vector of length 3 and/or a logical column index vector of length 4. For example,

```
> x[c(TRUE,TRUE,FALSE), c(FALSE,TRUE,FALSE,TRUE)]
     [,1] [,2]
[1,]    2    4
[2,]    6    8
> x[2,-3]
[1] 5 6 8
```

Finally, we show a method for selecting particular elements by use of an index-matrix. This is rarely needed, but useful when it is. The following example selects the three elements at (row, column) = (3,1), (2,2) and (1,3):

```
> mat.index = cbind(3:1, 1:3)
> mat.index
     [,1] [,2]
[1,]    3    1
[2,]    2    2
[3,]    1    3
```

```
> x[mat.index]
[1] 9 6 3
```

Notice that the index-matrix must have two columns, the first giving the row numbers and the second giving the corresponding column numbers.

4

Data handling

How do you select a subset of the data, combine two datasets, or remove an outlier? Skip this chapter at first reading, but look back into it when you have questions like these.

4.1 How to select parts of a data frame

For analyzing a subset of the data or for removing an outlier, for example, you may need this section.

As data frames are arranged in rows and columns, they are much like matrices, and you may use indexes with square brackets just as for matrices (except the version with an index-matrix). To select a subset according to a criterion, the `subset` function may, however, be more straightforward as shown next.

From the potato data frame in the introductory example, we may, for example, select the subset characterized by P81 being zero, by

```
> data0 = subset(potato, P81==0)
> data0
  P81 P82 yield
1   0   0   330
2   0  20   346
3   0  40   409
```

The same effect could be achieved by using logical indexing,

```
> data0 = potato[potato$P81==0,]
```

```
> data0
  P81 P82 yield
1   0   0   330
2   0  20   346
3   0  40   409
```

Notice, however, that in this command you cannot replace potato$P81 by P81 as you can with the subset function.

4.2 Combining data

Read this when you need to combine rows, columns or variables from different datasets.

To collect several data frames with the same variable names into a single data frame you use the function **rbind** (row bind). Suppose, for example, the potato data from the introductory example came as three data frames, data0, data30 and data60, with three lines of data each, and the first of which is

```
> data0
  P81 P82 yield
1   0   0   330
2   0  20   346
3   0  40   409
```

Then the data frame potato with nine data lines can be made by

```
> potato = rbind(data0, data30, data60)
> potato
  P81 P82 yield
1   0   0   330
2   0  20   346
3   0  40   409
4  30   0   368
5  30  20   371
6  30  40   409
7  60   0   360
8  60  20   382
9  60  40   398
```

To append extra columns (or variables) to a data frame you may use cbind similarly, but more likely you will construct new columns one by one as needed. For example,

```
> potato$ln.yield = log(potato$yield)
> head(potato, n=4)
  P81 P82 yield ln.yield
1   0   0   330 5.799093
2   0  20   346 5.846439
3   0  40   409 6.013715
4  30   0   368 5.908083
```

In more advanced applications you may want to combine two data frames with different information on the same individuals. For that purpose you use the function merge. The line

```
> merge(positions, diameters,
+       by=c("forest.id","tree.id"), all=TRUE)
```

is an example that merges the data frame positions of trees in a collection of forests with the data frame diameters, assuming that both data frames contain the variables forest.id and tree.id identifying the trees.

4.3 Analysis of data split into subgroups

Calculation of mean or standard deviation for each treatment group are examples of what you learn in this section.

Suppose you want to calculate the mean potato yield in the introductory example for the three subgroups given by the value of P81.

```
> by(potato$yield, INDICES=list(factor(potato$P81)),FUN=mean)
: 0
[1] 361.6667
------------------------------------------------------------
: 30
[1] 382.6667
------------------------------------------------------------
: 60
[1] 380
```

The third argument of by is the function that we want to apply on subgroups. The first argument is the vector that we want the calculations to be carried out on, and the `INDICES` argument must be a so-called `list` of factors. Here the list consists of just a single factor, namely the vector `potato$P81` which is converted to a factor by the function `factor`.

The same as above could be achieved by the function `aggregate`,

```
> a1 = aggregate(potato$yield, by=list(factor(potato$P81)),
+               FUN=mean)
> a1
  Group.1    x
1       0 361.6667
2      30 382.6667
3      60 380.0000
```

which is quite similar to `by`. For both of the functions you may save the results under a name, as shown above with `aggregate`.

If the analysis you want to carry out on each subgroup cannot be expressed as a single function name, you have two possibilities. The first possibility is to make a new dataset for each subgroup and then run the analyses one by one. The second possibility is to write your own function and use `by` with `FUN=` the name you have given the new function. How to write your function is beyond the present scope although it is not complicated.

4.4 Maintaining order

This section gives a few brief advices on how to organize your data within R. You may skip it at first reading.

When you are analyzing data it is often a long process involving several steps, such as analyzing parts of the data, removing an outlier, transforming variables and so on. Two advices are given here to keep order and overview.

- Keep related vectors in a data frame. Rather than using `attach` and working on the individual vectors it is easier to keep order if, for example, you remove and outlier by selecting a subset of the data frame in which the entire line is removed.

- Keep a script with the commands you use in each step. It is tempting sometimes to make the second step by making a few corrections in the first, but unless you are correcting errors, make a copy of

4.5 How to work on a data frame

This section indicates how to work on data frame, if you follow the advice not to attach it. You may skip it at first reading.

If you keep the data together in a data frame, as recommended in the previous section, without using `attach`, it remains slightly unpractical to have to write the name of the data frame in front of every single variable, like `potato$yield`.

There are basically three solutions you may choose. The first is to put up with the heavy but safe method and write these long names, but then take advantage of some functions ability to specify a data frame name just once, like in the introductory example, where

```
> lm1 = lm(yield~P82, data=potato)
```

uses the argument `data` in this way.

The second method is, after all, to use `attach` in a "hit and run" fashion: attach the data frame, do a single or a few steps, on the variables and close the session before you start working on modifications of the data — or at least "detach" the data frame with the `detach` function.

The third method is more professional and uses the functions `with` and `transform`. For example, the above function call with `lm` might have been written

```
> lm1 = with( potato, lm(yield~P82) )
```

Notice that you do not write

```
> with( potato, lm1 = lm(yield~P82) )   # WRONG
```

because the result `lm1` will not be available when the function `with` has terminated the calculations.

Similarly, if you want to make changes in the data frame, such as appending new columns, you may use `transform`,

```
> potato2 = transform(potato,
+                    ln.yield=log(yield),
+                    sqrt.yield=sqrt(yield))
> head(potato2, n=4)
  P81 P82 yield ln.yield sqrt.yield
1   0   0   330 5.799093   18.16590
2   0  20   346 5.846439   18.60108
3   0  40   409 6.013715   20.22375
4  30   0   368 5.908083   19.18333
```

As shown here, you may have several commands separated by commas. The indentations and the use of several lines make no difference and is only for ease of reading.

Part II

R for statistical analysis

Now we switch focus to the statistical analysis for which you will need a number of functions, some of which are described in the sequel. We assume that you are now prepared to work with R and know vaguely, at least, the concept and use of functions. From now on we drop the guiding lines telling what the chapter or section is needed for, as it should be more obvious here.

The datasets used are available in the package `Guide1data` which you should install and load as described in Appendix C to be able to read the data by writing `data(niacin)`, for example, as shown.

We remind you that there are hundreds of packages freely available for particular purposes or types of analysis, see Appendix A.

5
R functions for data presentation

In this short chapter we use some of the example datasets found in the package `Guide1data`, as just described. To run the examples you should therefore install that package and load it with the command `library(Guide1data)`. Next, to use the dataset `milkyield`, for example, you write `data(milkyield)`.

5.1 Functions listing data or data summaries

Simple statistics such as the sum or the mean may be calculated on a numeric vector by built-in functions. The vector we consider as example contains the 32 milk yields from the cows in group 1 in the `milkyield` dataset,

```
> data1 = subset(milkyield, Group==1)
> y1 = data1$Yield
> length(y1)
[1] 32
```

The following lines show the use of a number of functions applied to a vector.

```
> sum(y1)
[1] 150665
> mean(y1)
```

```
[1] 4708.281
> sd(y1)
[1] 648.1547
> var(y1)
[1] 420104.5
> median(y1)
[1] 4746
> min(y1)
[1] 3407
> max(y1)
[1] 5931
> which.min(y1)
[1] 12
> which.max(y1)
[1] 20
```

To explore a dataset we consider as an example the dataset `hydrolysis` from the package `Guide1data` described in Appendix C. The function `dim(hydrolysis)` will tell you that it has 50 rows and 3 columns, and

```
> data(hydrolysis)
> head(hydrolysis)
    feed hour serine
1 barley    8   4.47
2 barley   16   4.34
3 barley   24   4.22
4 barley   32   4.10
5 barley   72   3.48
6 barley    8   4.4
```

shows the top 6 lines. More comprehensive facts about the structure are obtained with the command `str(hydrolysis)`.

Finally, the function `table` is used to summarize categorical data, as it returns the count of each value. For example,

```
> table(hydrolysis$feed)
barley   fish   mais   meat    soy
    10     10     10     10     10
```

shows the five types of `feed` and that they occur ten times each, while

```
> table(hydrolysis$feed, hydrolysis$hour)
          8  16  24  32  72
barley    2   2   2   2   2
```

```
fish    2 2 2 2 2
mais    2 2 2 2 2
meat    2 2 2 2 2
soy     2 2 2 2 2
```

shows the 25 combinations of `feed` and `hour`, each appearing twice.

5.2 Graphical functions

Many basic types of plots are extremely simple and quick to make in R. For example,

```
> plot(x,y)      # Scatterplot of (x,y) for two equally long
+                # numeric vectors
> points(x,y)    # Adds points to the current plot
> lines(x,y)     # As points(x,y) but with points connected
+                # by line segments
> boxplot(x)     # Boxplot of the values i x
> hist(x)        # Histogram of the values in x
> pairs(data1)   # Matrix-scatterplot for all pairs in a data
+                # frame data1
> qqnorm(x)      # QQ-plot to check if x seems to follow a
+                # normal distribution
> interaction.plot(x.factor=treat, trace.factor=block,
+                  response=y)
+                # Interaction plot, see Figure 7.2
```

At the same time the graphical functions are very flexible; you can change almost anything in the plot: axes, labels, colors, sizes and so on, if you are not satisfied with the default choices. We show only a few graphical functions here and leave some for their particular use shown in subsequent sections. Instead we spend a little more effort on some useful options and utilities, many of which you can find if you use the help pages you get from

```
> ?par
> ?plot
```

and so on, where `par` is a special graphical function handling many graphical parameters. Notice also that other useful help pages may be found through the "See also" links.

It is worth noting at this point that R is object-oriented, meaning that functions may work differently (intelligently) depending on which object is the argument of the function. As an example, the function plot used in the form plot(x,y) produces a scatterplot if x and y are both numeric vectors, but produces parallel boxplots if x is a factor (a grouping variable, see Section 7.1.

5.2.1 The plot function

Consider the dataset hydrolysis in the Guide1data package, see Appendix C. Measurements of the concentration of the amino acid serine were made after different time of hydrolysis for different types of feed. The dataset has three variables: serine, feed and hour. A plot of serine against hour for the subset with barley bean can be made quickly by

```
> data(hydrolysis)
> hydro1 = subset(hydrolysis, feed=="barley")
> plot(hydro1$hour, hydro1$serine)
```

resulting in the left panel of Figure 5.1. Very often the default plot is quite satisfactory, but you might like to change details concerning the layout. The following call to plot shows some of the possibilities:

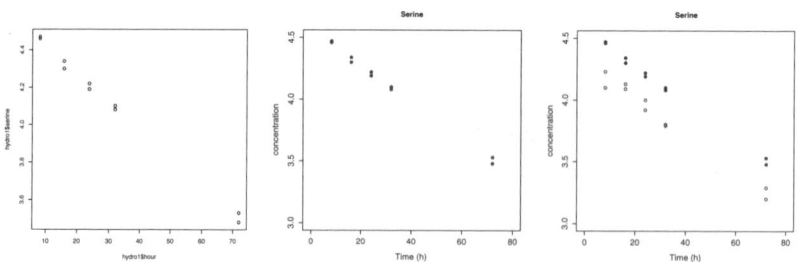

Figure 5.1. Scatterplots made by the plot function (see text).

```
> plot(hydro1$hour, hydro1$serine,
+       main="Serine",    # title above the plot
+       xlab="Time (h)", ylab="concentration", # text on axes
+       xlim=c(0,80),     # range of x-axis
+       ylim=c(3,4.5),    # range of y-axis
+       pch=16,           # the plotting character
+       col="brown",      # the plotting colour
+       cex=1.3,          # character extension factor for pch
```

5.2 Graphical functions

```
+         cex.lab=1.3,    # magnification factor for labels
+         cex.axis=1.3)   # and for values on axes
```

You see the result in the mid panel of Figure 5.1. The function `points` adds points to the current plot, so that

```
> hydro2 = subset(hydrolysis, feed=="fish")
> points(hydro2$hour, hydro2$serine, col="blue", pch=1)
```

results in the right panel of Figure 5.1 in which the "fish" results are shown as blue open circles. Notice how we extended the y axis in the `plot` command with `ylim` such that all points fall within the graph windoe.

The function `plot` has an option `type` which determines how you want the x and y coordinates to be presented. The default is that the points are shown with a plot symbol as above, whereas `type="l"` as in

```
> plot(x,y, type="l")
```

connects the points by line segments.

Similar to `points` you may add further lines to the current plot by writing

```
> lines(x,y, lty=2)
```

in which the optional argument `lty=2` specifies a dashed line, rather than the default solid line (1 is solid, 2 is dashed, 3 is dotted, etc.).

Often it is desirable to add a full line $y = a + bx$ to the current plot. This is exactly what is done by, for example,

```
> abline(a=2, b=3)
```

A horizontal and a vertical line can obtained using `abline(h=5)` or `abline(v=12)`, respectively.

5.2.2 How to add a legend

In the example from Figure 5.1 you would probably want to see all five types of feed in the same plot, and then also a legend showing which is which. There is nothing new in the commands

```
> hydro1 = subset(hydrolysis, feed=="barley")
> hydro2 = subset(hydrolysis, feed=="fish")
> hydro3 = subset(hydrolysis, feed=="mais")
> hydro4 = subset(hydrolysis, feed=="meat")
> hydro5 = subset(hydrolysis, feed=="soy")

> plot(hydro1$hour, hydro1$serine,
+       main="Serine",
+       xlab="Time (h)", ylab="concentration",
+       ylim=c(3,6),
+       pch=16,
+       col="brown",
+       cex=1.3, cex.lab=1.3, cex.axis=1.3)

> points(hydro2$hour, hydro2$serine, col="blue", pch=1)
> points(hydro3$hour, hydro3$serine, col="red", pch=2)
> points(hydro4$hour, hydro4$serine, col="black", pch=3)
> points(hydro5$hour, hydro5$serine, col="brown", pch=4)
```

which generate the plot in Figure 5.2, except for the legend at the upper right. This is obtained by the commands

```
> legend(x=60, y=6,
+        legend=c("barley","fish","mais","meat","soy"),
+        col=c("brown","blue","red","black", "brown"),
+        pch=c(16,1,2,3,4))
```

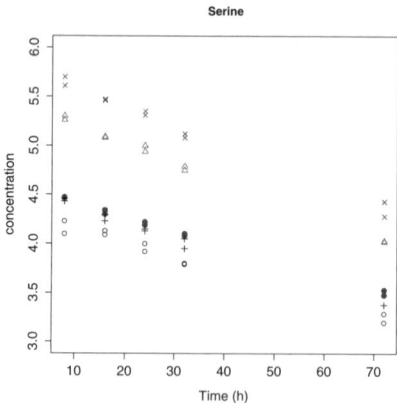

Figure 5.2. Plot with legend (see text).

If you want to place the legend visually you may instead use the function locator which waits for you to click on the plot and returns a pair of

5.2 Graphical functions

coordinates. In that case you write

```
> legend(locator(1),
+        legend=c("barley","fish","mais","meat","soy"),
+        col=c("brown","blue","red","black","brown"),
+        pch=c(16,1,2,3,4))
```

The argument in `locator(1)` means that you only want the location of one point.

There is also an argument `lty` to `legend` indicating the type of line you may have used to connect points.

5.2.3 How to save the plots

To save a plot the easiest way is to click the graph window and then use the menu (File, Save as, and so on), where you get a choice of many different file formats. The same can be programmed as, for example,

```
> savePlot("plot1", "png")
> savePlot("plot1", "pdf")
```

where the second argument is the file type and the first is the first part of the file name.

There is no plot history, but you can open an extra plot window by

```
> X11()
```

5.2.4 Further graphical utilities

In a graphical display it is sometimes desirable to mark or identify single points, for example an outlier. This is conveniently done by the function `identify`. For example,

```
> plot(x,y)
> identify(x,y,labels=patient)
```

will wait for you to click on points and mark them with the corresponding value of vector `patient`. If you omit this third argument, the points you click will be marked with their index number.

Finally let us mention the `lattice` package which has many useful (multi-display) graphical functions that are beyond the scope of this guide.

6
Basic statistical functions and tests

In this chapter we illustrate the use of some R functions for probability distributions and some well-known classical statistical tests, including some non-parametric tests.

We first introduce the normal distribution and show how to perform some of the classical tests based on an assumption of normality such as the t-test. Then we proceed to discrete distributions and tests based on those.

Remember that to use the datasets from the package Guide1data, it is necessary to install and load that package first, see Appendix C.

6.1   The normal distribution

The density function, distribution function and the random number generator function of the normal distribution are called dnorm, pnorm and rnorm, respectively. If X is assumed to be normally distributed with mean 4200 and standard deviation 420 (variance 420^2) then we can calculate $P(X \leq 4500)$, say, using pnorm,

```
> pnorm(4500, mean=4200, sd=420)
[1] 0.7624747
```

and the right-tail probability, $P(X > 4500)$, as

```
> 1-pnorm(4500, mean=4200, sd=420)
[1] 0.2375253
```

The mean and standard deviation are optional arguments and when leaving them out they are set to 0 and 1, respectively. The last computation above could also have been performed by

```
> pnorm(4500, mean=4200, sd=420, lower.tail=FALSE)
[1] 0.2375253
```

Quantiles of the standard normal distribution can be obtained using the function qnorm as shown below

```
> qnorm(0.5)      # median
[1] 0
> qnorm(0.975)    # 97.5% quantile
[1] 1.959964
```

As above, you can specify the mean and standard deviation with mean and sd.

Random numbers from the normal distribution are generated with rnorm. For example,

```
> rnorm(n=6, mean=15, sd=2)
[1] 15.22621 13.28919 12.78741 13.19356 15.86115 14.04146
```

gives six random numbers from the normal distribution with mean 15 and standard deviation 2 (variance 4).

Other distributions. In R the system of names with "d" for density, "p" for probability distribution function, "q" for quantile, and "r" for random sample is used also for other distributions. For example dt, pt, qt and rt are the corresponding names for the t-distribution. In the t-distribution the quantiles qt are frequently used, see Section 6.2, whereas you will hardly ever need dt or rt.

Normal quantile plot. A normal quantile plot is obtained using qqnorm. We use the milk yield data from the package Guide1data for illustration, see Appendix C. In particular you may write ?milkyield in R for info on these data

6.2 Parametric two-sample tests

```
> data(milkyield)
> data1 = subset(milkyield, Group==1)
> qqnorm(data1$Yield)
> qqline(data1$Yield)   # Adds a line to the plot
```

Kolmogorov-Smirnov, Shapiro-Wilks tests. The classical Kolmogorov-Smirnov test and the Shapiro-Wilks test for normality are obtained as

```
> ks.test(data1$Yield, "pnorm", mean=4708, sd=648)

        One-sample Kolmogorov-Smirnov test

data:  data1$Yield
D = 0.0833, p-value = 0.9794
alternative hypothesis: two.sided

> shapiro.test(data1$Yield)

        Shapiro-Wilk normality test

data:  data1$Yield
W = 0.9818, p-value = 0.85
```

showing in this case no sign of deviation from the normal distribution.

6.2 Parametric two-sample tests

Paired t-test. Consider the data from the package Guide1data, see Appendix C, concerning the effect of niacin with respect to the concentration of hemoglobin in the blood of some dogs. The concentration of hemoglobin was measured for each dog before and after treatment.

```
> data(niacin)
> niacin
  dog  pre after
1   1 12.6  10.4
2   2 12.6  11.5
3   3 13.7  13.6
4   4 11.1  12.0
5   5 11.3  10.7
6   6 12.2   9.3
```

```
7    7 10.0   8.8
8    8 11.4   9.4
```

```
> t.test(niacin$pre, niacin$after, paired=TRUE,
+          conf.level=0.95)
```

```
Paired t-test

data:  niacin$pre and niacin$after
t = 2.6558, df = 7, p-value = 0.03267
alternative hypothesis: true difference in means is not
      equal to 0
95 percent confidence interval:
 0.1260877 2.1739123
sample estimates:
mean of the differences
                   1.15
```

We see that the paired t-test is calculated and the p-value of 0.03 indicates an effect of niacin on the concentration of hemoglobin for the dogs considered. The `t.test` function calculates the t-test statistic (2.6558) and the corresponding p-value (0.03267). Let us try to calculate the p-value directly keeping in mind that the t-test in this case is two-sided. The p-value is therefore given as

```
> 2*(1-pt(2.6558, df=7))
[1] 0.03266557
```

or

```
> 2*pt(2.6558, df=7, lower.tail=FALSE)
[1] 0.03266557
```

t-test for comparisons of means of independent samples. We use the milk yield data to illustrate how to do the un-paired t-test.

```
> data(milkyield)
> data1 = subset(milkyield, Group==1)
> data2 = subset(milkyield, Group==2)
> t.test(data1$Yield, data2$Yield, var.equal=TRUE)
```

```
Two Sample t-test
```

6.2 Parametric two-sample tests

```
data:  data1$Yield and data2$Yield
t = -3.8386, df = 63, p-value = 0.0002894
alternative hypothesis: true difference in means is
        not equal to 0
95 percent confidence interval:
 -924.6455 -291.5192
sample estimates:
mean of x mean of y
 4708.281  5316.364
```

We see that the t-test value and associated p-value are reported, and the 95% confidence interval for the mean differences is also given. To do the t-test without assuming equal variances, the var.equal argument is set to FALSE (or omitted since this is the default):

```
> t.test(data1$Yield, data2$Yield)

        Welch Two Sample t-test

data:  data1$Yield and data2$Yield
t = -3.8368, df = 62.765, p-value = 0.0002919
alternative hypothesis: true difference in means is
          not equal to 0
95 percent confidence interval:
 -924.8170 -291.3478
sample estimates:
mean of x mean of y
 4708.281  5316.364
> # Equivalent:
+ # t.test(data1$Yield, data2$Yield, var.equal=FALSE)
```

giving almost the same as the classical t-test. This is so because the assumption about equal variation in the two groups is reasonable here. As a matter of fact we can investigate that by use of the Bartlett test:

```
> bartlett.test(Yield~Group, data=milkyield)

        Bartlett test of homogeneity of variances

data:  Yield by Group
Bartlett's K-squared = 0.0278, df = 1, p-value = 0.8675
```

confirming that the assumption of equal variances is reasonable.

6.3 Discrete distributions

The binomial distribution. The binomial distribution with parameters (n, p) has probability function (sometimes called density function)

$$f(x) = \binom{n}{x} p^x (1-p)^{n-x}, \quad \text{for } x = 0, \ldots, n.$$

We may use R to calculate values of the density function and of the corresponding distribution function using dbinom and pbinom, respectively. Assume that $n = 5$, $p = 1/3$. Let us calculate the density function in 0, i.e. $f(0)$, and then in all possible x-values, and finally take the sum of the density values (needs to give 1!).

```
> dbinom(0,size=5,prob=1/3)
[1] 0.1316872
> dbinom(0:5,size=5,prob=1/3)
[1] 0.131687243 0.329218107 0.329218107 0.164609053
[5] 0.041152263 0.004115226
> sum(dbinom(0:5,size=5,prob=1/3))
[1] 1
```

The distribution function (which consists of the successively cumulated probabilities) in 0 and in all the x-values is calculated as

```
> pbinom(0,size=5,prob=1/3)
[1] 0.1316872
> pbinom(0:5,size=5,prob=1/3)
[1] 0.1316872 0.4609053 0.7901235 0.9547325 0.9958848
[6] 1.0000000
```

and we may draw observations (20, say) at random from the distribution by

```
> rbinom(20,size=5,prob=1/3)
[1] 4 2 0 1 2 3 1 1 0 0 1 1 2 1 0 1 1 2 3 1
```

Drawing a very large sample size and calculating the empirical mean should give us a value close to the theoretical mean which is np:

```
> X = rbinom(200000,size=5,prob=1/3)
> mean(X)
[1] 1.666945
> 5/3
[1] 1.666667
```

6.3 Discrete distributions

This is indeed the case. The names of the functions dbinom (density function), pbinom (probability distribution function) and rbinom (random sample function) follow the convention mentioned in Section 6.1.

The multinomial distribution. The multinomial distribution extends the binomial distribution to the case with more than two outcomes for each trial. Let us consider the situation where we have 4 possible outcomes (such as a scoring of how affected a given plant is of certain disease). Further assume that the true probability parameters are 0.49,0.23,0.04,0.23. We may then generate a sample from this multinomial distribution with $n = 142$, say, as

```
> rmultinom(1, 142, prob=c(0.49,0.23,0.04,0.23))
     [,1]
[1,]   64
[2,]   38
[3,]    4
[4,]   36
```

and also calculate the density function in a given point such as for example ($X_1 = 70, X_2 = 33, X_3 = 6, X_4 = 33$):

```
> dmultinom(c(70,33,6,33),size=142,
+           prob=c(0.49,0.23,0.04,0.23))
[1] 0.001090018
```

The Poisson distribution. In an experiment with 150 leaves from apple trees one counted the number of a certain mite (European red mite) on each leave (Bliss, 1953). The results are given in the below table.

Number of mites per leave	0	1	2	3	4	5	6	7	≥ 8
Number of leaves	70	38	17	10	9	3	2	1	0

Such data are called count data, and one distribution that might be appropriate to describe such data is the Poisson distribution. It has probability function

$$P(X = x) = \frac{\lambda^x}{x!} e^{-\lambda}, \quad x = 0, 1, \ldots.$$

for a fixed $\lambda > 0$. If we use the Poisson distribution to describe these data, then the estimate of λ is 1.1467 identical to the sample mean. To check whether this Poisson distribution fits the dataset we may use the following R code to produce Figure 6.1.

```
> x = 0:7
> N = c(70,38,17,10,9,3,2,1)
> lambda = sum(x*N)/sum(N)
> lambda
[1] 1.146667
> pred = sum(N) * dpois(x,lambda)
> pred
[1] 47.65409795 54.64336565 31.32886297 11.97458762
[5]  3.43271512  0.78723600  0.15044955  0.02464507
> plot(x, N, type='h')
> lines(x+0.2, pred, type='h', lty=2)
> legend(x=4, y=60, legend=c("Observed","Fitted"), lty=1:2)
```

As illustrated above the density function of the Poisson distribution is dpois. R also supports the distribution function and a random number generator based on the Poisson distribution, try ?dpois to learn more about these. As is clear from Figure 6.1, the Poisson distribution does not give a good a fit to these data. One explanation might be that the leaves were collected from in total 25 trees with 6 leaves per tree so there may be more variation in the data than can be accounted for by the Poisson distribution. A better model in this case is based on the negative binomial distribution which is also available in R, see for example ?dnbinom.

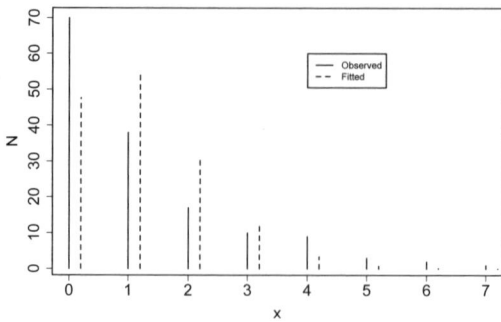

Figure 6.1. Mites on apple leaves. Observed and fitted data based on the Poisson distribution

6.4   Tests in contingency tables

Test of simple hypothesis in the multinomial distribution. Consider the following data on fungal spores (Penicillium Sp.) measured over the year in a specific household in Denmark; the numbers given are the number of spores.

Jan	Feb	Mar	Apr	May	Jun	Jul	Aug	Sep	Oct	Nov	Dec
14	26	16	12	10	17	10	20	15	17	14	8

If N_i denotes the number spores in the ith month then $(N_1, \ldots, N_{12})$ has a multinomial distribution with $n = 179$ (total number of fungal spores over the year) and with probability vector $(p_1, \ldots, p_{12})$. We can test the null hypothesis $p_1 = \cdots = p_{12} = 1/12$ in R using

```
> N = c(14,26,16,12,10,17,10,20,15,17,14,8)
> chisq.test(N, p=rep(1/12,12))

        Chi-squared test for given probabilities

data:  N
X-squared = 17.7598, df = 11, p-value = 0.08732
```

The argument p=rep(1/12,12) states that we are testing the hypothesis that $p_i = 1/12$, $i = 1, \ldots, 12$, which is, in fact, default and thus could have been omitted. We see that the test is not significant at the 5% significance level so we cannot rule out the null hypothesis.

Test of homogeneity, the chi-square test. The chi-square test of homogeneity is performed using the chisq.test function again. Consider the following data on nest predation where 85 artificial nests were placed at a varying density of reeds.[1] After some time it was registered whether or not the nests had been looted, and the question is whether the density of reeds gives some protection against the nest being looted.

The test statistic and corresponding p-value is calculated as follows:

```
> X = matrix(c(22,6,23,6,14,14),nrow=3,ncol=2,byrow=TRUE)
```

---

[1] The data come from V. Salonen and A. Penttinen (1988), "Factors affecting nest predation in the Great Crested Grebe: field observation, experiments and their statistical analysis", *Ornis Fennica*, **65**.

	Density of reed (reeds/m^2)		
	10 – 40	50 – 80	≥ 90
Nest being looted, yes	22	23	14
Nest being looted, no	6	6	14

```
> X
     [,1] [,2]
[1,]  22    6
[2,]  23    6
[3,]  14   14

> chisq.test(X)

        Pearson's Chi-squared test

data:  X
X-squared = 7.4141, df = 2, p-value = 0.02455

> p.hat = c(22/28,23/29,14/28)
> p.hat
[1] 0.7857143 0.7931034 0.5000000
```

The hypothesis of homogeneity is thus rejected at the conventional 5% significance level with $p = 0.02$. The function prop.test also calculates the test statistic and in addition provides the estimates of the probability parameters:

```
> prop.test(c(22,23,14), n=c(28,29,28))
```

If we use prop.test only on two groups it will also provide a 95% confidence interval of $p_1 - p_2$:

```
> prop.test(c(22,14), n=c(28,28))

        2-sample test for equality of proportions with
        continuity correction

data:  c(22, 14) out of c(28, 28)
X-squared = 3.8111, df = 1, p-value = 0.05091
alternative hypothesis: two.sided
95 percent confidence interval:
 0.01042143 0.56100714
sample estimates:
```

6.4 Tests in contingency tables

```
      prop 1    prop 2
   0.7857143 0.5000000
```

comparing here group 1 and 3 in the example.

Fisher's exact test. Consider the following data on a possible effect of a fungicide, avadex, on the risk of developing tumor in the lungs. A group of mice (16) were given avadex in their feed while another group of mice (79) were given the standard feed. It was observed whether the mice developed tumor.

	With avadex	Without avadex
Tumor, yes	4	5
Tumor, no	12	74

The chi-square test statistic and corresponding p-value is calculated as usual

```
> X = matrix(c(4,12,5,74), nrow=2, ncol=2, byrow=TRUE)
> X
     [,1] [,2]
[1,]    4   12
[2,]    5   74
> fit = chisq.test(X)
Warning message:
In chisq.test(X) : Chi-squared approximation may be incorrect

> fit$expected
         [,1]     [,2]
[1,] 1.515789 14.48421
[2,] 7.484211 71.51579
> fit

    Pearson's Chi-squared test with Yates' continuity correction

data:  X
X-squared = 3.4503, df = 1, p-value = 0.06324
```

We see that one of the expected counts is less than 5 and we indeed get a warning that the chi-square test might be in-appropriate to use for these data. Instead we can use the Fisher exact test:

```
> fisher.test(X)
```

```
        Fisher's Exact Test for Count Data

data:  X
p-value = 0.04106
alternative hypothesis: true odds ratio is not equal to 1
95 percent confidence interval:
  0.834087 26.162982
sample estimates:
odds ratio
  4.814787
```

which rejects the null hypothesis of homogeneity.

McNemar's test. In the case where we have a binary response and wish to compare two groups (treatments) based on paired observations we may use the McNemar's test. For instance, in a 2x2 crossover trial on cerebrovascular deficiency each of 67 patients was subjected to an active treatment and placebo over two consecutive periods. In each period, the patient's electrocardiogram (ec) was categorized as either normal or abnormal (the response). The data from this study are given in the below table.

	Active treatment	
Placebo	abnormal ec	normal ec
abnormal ec	15	10
normal ec	2	40

We see that 20 patients had a normal electrocardiogram when treated while their electrocardiogram was judged abnormal when untreated. The data are actually an example of repeated binary data, and we show in Section 10.1 how to deal with such data in more generality. In the situation with only two time points of observation, we can use McNemar's test as follows (ignoring a possible difference between the two time periods):

```
> X = matrix(c(15,2,10,40), nrow=2, ncol=2)
> X
     [,1] [,2]
[1,]   15   10
[2,]    2   40
> mcnemar.test(X)

        McNemar's Chi-squared test with continuity correction
```

```
data:  X
McNemar's chi-squared = 4.0833, df = 1, p-value = 0.04331
```

The test shows that there seems to be a significant treatment effect, $p = 0.04$. This would have been overlooked if we wrongly ignored the pairing and just performed an ordinary chi-square test:

```
> X = matrix(c(25,17,42,50), nrow=2, ncol=2)
> X
     [,1] [,2]
[1,]   25   42
[2,]   17   50
> chisq.test(X)   # Not appropriate
Pearson's Chi-squared test with Yates' continuity correction
X-squared = 1.6993, df = 1, p-value = 0.1924
```

Notice that the hypothesis tested by the ordinary chi-square test is that row classification is independent of column classification, whereas McNemar's test concerns the hypothesis that row classes have the same probabilities as the column classes. Thus, the hypotheses being tested by the two methods are widely different.

The Kappa statistic. Sometimes it is of interest to measure the degree of agreement between two methods. This may be evaluated using the Kappa statistic as shown in the following example concerning the classification by two radiologists of 85 xeromammograms as normal, benign disease, suspected cancer, and cancer: In this case there is agree-

Radiologist A's assessment	Radiologist B's assessment			
	normal	benign	suspect	cancer
normal	21	12	0	0
benign	4	17	1	0
suspect	3	9	15	2
cancer	0	0	0	1

ment in 54 of 85 cases (64%). Is this satisfactory, one may ask, and how much agreement can be expected just by chance? If there is no relation at all, then we would expect $\frac{28}{85} \cdot \frac{33}{85} \cdot 85 = 10.87$ observations in the first diagonal entry. Similarly we would expect 9.84, 5.46 and 0.04, respectively, in the three other diagonal entries, and thus 26.20 in total on the diagonal. Hence, the expected agreement just by chance is 26.20/85 or 31%. Since the maximum agreement is 1 (agree on all cases) the possible scope for doing better than chance is this example $1 - 0.31$. The Kappa statistic is given by $(0.64 - 0.31)/(1 - 0.31) = 0.47$, measuring how much we are doing better than chance as a proportion of the possible scope for doing better than chance. To calculate in the Kappa statistic R you do as follows:

```
> library(vcd)
> x = matrix(c(21,12,0,0,4,17,1,0,3,9,15,2,0,0,0,1),
+            nrow=4, ncol=4, byrow=T)
> x
     [,1] [,2] [,3] [,4]
[1,]   21   12    0    0
[2,]    4   17    1    0
[3,]    3    9   15    2
[4,]    0    0    0    1
> Kappa(x)
              value        ASE
Unweighted 0.4727891 0.07547288
Weighted   0.5683990 0.12536834
```

R also reports a weighted version of the Kappa statistic taking off diagonal entries into account, but we will not go into details with this value. Usually one uses the following classification concerning strength of agreement for the obtained value of Kappa:

Value of Kappa				
<0.20	0.21–0.40	0.40–0.60	0.60–0.80	0.80–1.00
Poor	Fair	Moderate	Good	Very good

so in the example we conclude that we have moderate agreement.

6.5 Non-parametric tests

Wilcoxon's rank and signed rank test. Sometimes the assumptions behind using the t-test fail and it may not be possible to find a transformation of the response so that model requirements are met. In such cases one may apply the Wilcoxon rank sum test . This is a non-parametric test, i.e., it is neither based on any assumption concerning the normal distribution nor variance homogeneity. In the paired case it

6.5 Non-parametric tests

is called the Wilcoxon signed rank test. For illustration we use the data also used for illustration of the t-test, see Section 6.2. First the paired situation:

```
> data(niacin)
> wilcox.test(niacin$pre, niacin$after, paired=TRUE)

        Wilcoxon signed rank test

data:  niacin$pre and niacin$after
V = 33, p-value = 0.03906
alternative hypothesis: true location shift is not equal to 0
```

and then the un-paired situation:

```
> data(milkyield)
> wilcox.test(Yield~Group, data=milkyield)

        Wilcoxon rank sum test

data:  Yield by Group
W = 263, p-value = 0.0003816
alternative hypothesis: true location shift is not equal to 0
```

In both situations we obtain p-values very close to those obtained by the corresponding t-tests, which is because model requirements are fulfilled in these two situations.

To see that there may indeed be a difference between the parametric and non-parametric test, we turn to an example concerning streptococci.[2] (Altman, 1990). The data are saved as the data frame streptococcus in the Guide1data, see Appendix C:

```
> data(streptococcus)
> head(streptococcus)
  id  bi  ai
1  1 0.4 0.4
2  2 0.4 0.5
3  3 0.4 0.5
4  4 0.4 0.9
5  5 0.5 0.5
6  6 0.5 0.5
```

---

[2] The data come from D.G. Altman (1990), *Practical Statistics for Medical Research*, Chapman & Hall/CRC.

The variables `bi` and `ai` give the concentrations of antibody to Type III Group B Streptococcus before and after immunization, and `id` is an identification number for the individuals. Twenty individuals participated.

Let us do the paired t-test and its corresponding non-parametric variant:

```
> t.test(streptococcus$bi, streptococcus$ai, paired=TRUE,
+         conf.level=0.95)

Paired t-test

data:  streptococcus$bi and streptococcus$ai
t = -1.8498, df = 19, p-value = 0.07996
alternative hypothesis: true difference in means is
        not equal to 0
95 percent confidence interval:
 -2.5151563  0.1551563

> wilcox.test(streptococcus$bi, streptococcus$ai,
+              paired=TRUE)

Wilcoxon signed rank test with continuity correction

data:  streptococcus$bi and streptococcus$ai
V = 2, p-value = 0.00412
alternative hypothesis: true location shift is
        not equal to 0
```

Here we see a marked difference between the two reported p-values. Inspecting the observations it is clear that the assumptions behind the paired t-test cannot be fulfilled, try also the following:

```
> streptococcus$diff = streptococcus$ai-streptococcus$bi
> qqnorm(streptococcus$diff)
```

Kruskal-Wallis test. If you have more than two groups to be compared corresponding to the one-way ANOVA (see Section 7.1.1) but it does not seem reasonable to apply a linear normal model, then there is also a non-parametric test for this situation. It is called the Kruskal-Wallis test. We use a dataset on chlorophyll production in winter wheat as illustration, the dataset is part of the `Guide1data` package.

```
> data(chlorophyll)
> chlorophyll
```

```
   treat chloro
1     1     54
2     1     46
3     1     44
4     1     53
5     2     62
6     2     65
7     2     74
8     2     68
9     3     76
10    3     69
11    3     84
12    3     74
> kruskal.test(chloro~treat, data=chlorophyll)
        Kruskal-Wallis rank sum test

data:  chloro by treat
Kruskal-Wallis chi-squared = 9.0412, df = 2, p-value = 0.0109
```

so we find a significant treatment effect based on this non-parametric test, compare to the one-way ANOVA in Section 7.1.1.

6.6 Power and sample size calculations

An important part of the planning of a statistical study is to decide how many units (people, animals, plants, etc.) to include in the study. If the study is small, it may be difficult to find true effects. Therefore a formal sample size calculation should be carried out before starting collecting data. Such a calculation needs some input from the investigator before it can be carried out, however. There are some built-in functions in R to do such calculations, but unfortunately they can only deal with rather simple situations (described below). In more complicated situations, you would have to computed the power or sample size by simulation.

Continuous response, two groups. Say we have two groups to be treated and that the response to be measured is continuous. We wish to compute the sample size that with a power of 80% (using the two-sided t-test at the 5% level) can find a difference of 10 in a distribution with standard deviation of 15, say. This can be calculated using `power.t.test` function:

```
> power.t.test(delta=10, sd=15, sig.level=0.05, power=0.8)
```

```
                Two-sample t test power calculation

                     n = 36.3058
                 delta = 10
                    sd = 15
             sig.level = 0.05
                 power = 0.8
           alternative = two.sided
       NOTE: n is number in *each* group
```

Hence, we need in total 72 units. You can also compute the power by giving the sample size. If for example we can only afford to include 40 units, then the following shows that we only have a power of around 50%:

```
> power.t.test(n=20, delta=10, sd=15, sig.level=0.05)

                Two-sample t test power calculation

                     n = 20
                 delta = 10
                    sd = 15
             sig.level = 0.05
                 power = 0.5377573
           alternative = two.sided
       NOTE: n is number in *each* group
```

Continuous response, more than two groups. With more than two groups we can use the function pwr.anova.test in the package pwr. The effect size is given as

$$f = \sqrt{\sum_{i=1}^{k} \frac{(\mu_i - \mu)^2 n_i/N}{\sigma^2}},$$

with values 0.1, 0.25 and 0.4 corresponding to small, medium and large effect sizes, respectively. In the latter display k is the number of groups, μ_i is the mean of group i, μ is the grand mean, n_i is number of observations in group i, N is the total number of observations and σ^2 is the variance within groups (assumed to be the same for all groups). The call to the function is as follows

6.6 Power and sample size calculations

```
> library(pwr)
> pwr.anova.test(k=3, f=0.25, sig.level=0.05, power=0.8)

     Balanced one-way analysis of variance power calculation

              k = 3
              n = 52.3966
              f = 0.25
      sig.level = 0.05
          power = 0.8

NOTE: n is number in each group
```

There is also the possibility of giving the sample size and then the function calculates the power.

Binary response, two groups. If instead we are planning a study where the response is binary then it is the function power.prop.test that should be used. Say we want to compute the sample size that with a power of 90% can find a difference corresponding to the two proportions $p_1 = 0.25$ and $p_2 = 0.4$:

```
> power.prop.test(power=0.9, p1=0.25, p2=0.40)

     Two-sample comparison of proportions power calculation

              n = 202.8095
             p1 = 0.25
             p2 = 0.4
      sig.level = 0.05
          power = 0.9
    alternative = two.sided
NOTE: n is number in *each* group
```

It turns out that we need a bit more than 200 observations in each of the two groups.

7
Linear normal models

Linear models comprise analysis of variance, linear regression and much more. They are fitted with the `lm` function. A call to `lm` produces an object which can then be further inspected with various other functions. For example, `summary` gives us parameter estimates as well as their standard errors and the corresponding t-tests; `confint` produces confidence intervals for the parameters; and `anova` gives the analysis of variance table and can be used for testing nested linear models against each other.

7.1 Analysis of variance (ANOVA)

In this section we consider the class of ANOVA models, that is, models where all the explanatory variables are factors. First we analyse one-way ANOVA models, then the two-way (and multi-way) ANOVA models.

7.1.1 One-way ANOVA

Let us consider a dataset on chlorophyll production in winter wheat. We first read in the data from the package `Guide1data`, see Appendix C. Then we plot `chloro` against `treat` with the `plot` command and do a boxplot with the `boxplot` command. The plots are shown in Figure 7.1.

```
> data(chlorophyll)
> plot(chloro~treat,
     data=chlorophyll)    #Scatterplot, Fig. 7.1 (left)
> boxplot(chloro~treat,
     data=chlorophyll)    #Boxplots, Fig. 7.1 (right)
```

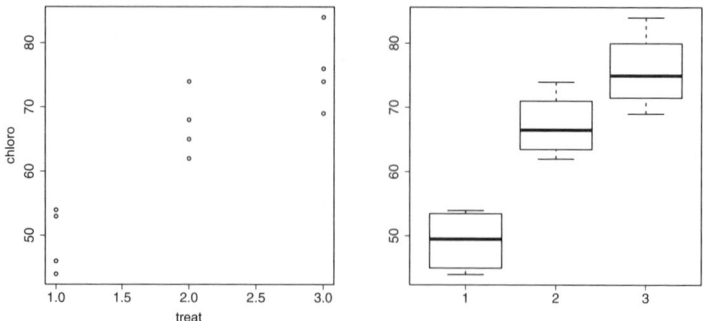

Figure 7.1. Scatterplot and boxplot of the Chlorophyll data.

Model fit, analysis of variance table, test for treatment effect.
The one-way ANOVA model

$$Y = \alpha(\text{treat}) + e$$

is fitted with lm. The response (chloro) is written on the left hand side of a ~ (a "tilde"), the factor (treat) on the right hand side. We need to tell R to use treat as a factor rather than as a numerical variable; we do so with factor:

```
> chlorophyll$treat = factor(chlorophyll$treat)
> model1 = lm(chloro~treat, data=chlorophyll)
```

We could also have used the factor function inside the lm call and fitted the model with

```
> lm(chloro~factor(treat), data=chlorophyll)
```

The analysis of variance table is obtained with anova:

```
> anova(model1)
Analysis of Variance Table

Response: chloro
          Df  Sum Sq Mean Sq F value    Pr(>F)
treat      2 1464.67  732.33  24.389 0.0002324 ***
Residuals  9  270.25   30.03
```

From the table we read the residual sum of squares (SSe=270.25) the residual degrees of freedom (DFe=9) as well as the residual mean square

7.1 Analysis of variance (ANOVA)

error (MSe=SSe/DFe=30.03). Moreover, we see that there is a highly significant treatment effect (p=0.00023). The test for treatment effect could also be carried out by fitting the model without treatment effect and compare it to the model including the treatment effect using the anova function as follows:

```
> model2 = lm(chloro~1, data=chlorophyll) # No treatm. effect
> anova(model2, model1)  # Test ag. model with treatm. effect
Analysis of Variance Table

Model 1: chloro ~ 1
Model 2: chloro ~ treat
  Res.Df    RSS Df Sum of Sq      F    Pr(>F)
1     11 1734.92
2      9  270.25  2    1464.67 24.389 0.0002324 ***
```

Parameter estimates. The parameter estimates are extracted with summary. The corresponding standard errors are also listed, and the t-tests for each parameter being zero (one at a time) are reported. Some of the output has been suppressed for clarity:

```
> summary(model1)

Coefficients:
            Estimate Std. Error t value Pr(>|t|)
(Intercept)   49.250      2.740  17.975 2.32e-08 ***
treat2        18.000      3.875   4.645  0.00121 **
treat3        26.500      3.875   6.839 7.57e-05 ***

Residual standard error: 5.48 on 9 degrees of freedom
Multiple R-Squared: 0.8442,     Adjusted R-squared: 0.8096
F-statistic: 24.39 on 2 and 9 DF,  p-value: 0.0002324
```

We see that $\hat{\sigma} = 5.48$ (residual standard error). For the treatment effect, R uses the first level of treat as reference. That is, the first number in the Estimate column is the estimate of $\alpha(1)$: $\hat{\alpha}(1) = 49.25$. Then follows the estimate $\hat{\alpha}(2) - \hat{\alpha}(1) = 18.00$ and $\hat{\alpha}(3) - \hat{\alpha}(1) = 26.50$. Hence, in order to get an estimate of $\alpha(2)$, we compute

$$\hat{\alpha}(2) = \hat{\alpha}(1) + (\hat{\alpha}(2) - \hat{\alpha}(1)) = 49.25 + 18.00 = 67.25.$$

A slightly different way of thinking about the estimates is the following. Think about the model written with an intercept, μ:

$$Y_i = \mu + \beta(\text{treat}) + e_i$$

and fix $\beta(1)$ to zero. With this parametrization, $\hat{\mu} = 49.25$, $\hat{\beta}(2) = 18.00$, $\hat{\beta}(3) = 26.50$.

Note that we could have got the α-estimates right away by fitting the model "without intercept":

```
> model1a = lm(chloro~treat-1, data=chlorophyll)
> summary(model1a)
Coefficients:
        Estimate Std. Error t value Pr(>|t|)
treat1    49.25       2.74   17.98 2.32e-08 ***
treat2    67.25       2.74   24.55 1.48e-09 ***
treat3    75.75       2.74   27.65 5.14e-10 ***
Residual standard error: 5.48 on 9 degrees of freedom
Multiple R-squared: 0.9947, Adjusted R-squared: 0.9929
F-statistic: 563.3 on 3 and 9 DF,  p-value: 1.480e-10
```

It is important to realize that `model1` and `model1a` are exactly the same; the only difference is the parameterization. In particular, of course, the two models give the same residual standard error.

Confidence intervals. Confidence intervals are easily obtained with `confint`. For the original model specification with group 1 as reference, we get

```
> confint(model1)
                2.5 %    97.5 %
(Intercept) 43.051960 55.44804
treat2       9.234648 26.76535
treat3      17.734648 35.26535
```

Change of reference group. By default, the first treatment group is used a reference, but the reference group can be changed with `relevel`:

```
> chlorophyll$newtreat = relevel(factor(chlorophyll$treat),
+                                ref=2)
> model1b = lm(chloro~newtreat, data=chlorophyll)
> summary(model1b)
Coefficients:
            Estimate Std. Error t value Pr(>|t|)
(Intercept)   67.250      2.740  24.545 1.48e-09 ***
newtreat1    -18.000      3.875  -4.645  0.00121 **
```

7.1 Analysis of variance (ANOVA)

```
newtreat3      8.500      3.875   2.194  0.05591 .

Residual standard error: 5.48 on 9 degrees of freedom
Multiple R-squared: 0.8442,Adjusted R-squared: 0.8096
F-statistic: 24.39 on 2 and 9 DF,  p-value: 0.0002324
```

In particular we see that the p-value for the hypothesis $\alpha(2) = \alpha(3)$ is just above 5%.

7.1.2 Two-way ANOVA, multi-way ANOVA

In this section we will use data concerning decomposition of organic matter.[1] A commonly used veterinary medicine (Ivermectin) was investigated for its influence on the decomposition of dung organic matter from treated cattle. Dung was collected two days after treatment and mesh bags with portions of 40 g of dung was placed in the soil. The organic matter was then determined in 6 mesh bags retrieved after 8, 12 and 16 weeks. The influence of the veterinary medicine was investigated by comparing with the decomposition of dung matter from untreated cattle.

First, load the data into R from the package `Guide1data`, see Appendix C. Next, make the time-variable, `tim`, a factor. The medicine factor, `vet`, is automatically a factor since its levels are not numeric.

```
> data(organic)
> organic$tim = factor(organic$tim)
```

Interaction, fit of model. First, some general comments: The syntax for the product of two factors, A and B, is `A:B`, whereas `A*B` is short for the collection of `A:B`, A, B and 0 (the trivial factor). Hence `model1`, `model1a` and `model1b` below all fit the two-way ANOVA model with interaction

```
> model1 = lm(y ~ A + B + A:B)
> model1a = lm(y ~ A*B)
> model1b = lm(y ~ A:B)
```

The fits `model1` and `model1a` are exactly the same, whereas `model1b` fits the same model but with another parameterization. The latter might

---

[1] The experiment was conducted by Christian Sommer, Department of Ecology, University of Copenhagen as part of a larger study.

be useful for estimation purposes but for hypothesis testing model1 or model1a is generally recommended. In particular, the output from anova(model1b) is almost never useful.

For the example, the model fit and the analysis of variance table goes as follows:

```
> model1 = lm(matter~vet+tim+vet:tim, data=organic)
> anova(model1)
Analysis of Variance Table

Response: matter
          Df  Sum Sq Mean Sq F value    Pr(>F)
vet        1 1.73155 1.73155 79.0972 6.519e-10 ***
tim        2 0.76809 0.38405 17.5432 9.004e-06 ***
vet:tim    2 0.00767 0.00383  0.1751    0.8402
Residuals 30 0.65674 0.02189
```

From the analysis of variance table we get the variance estimate, $\hat{\sigma}^2 = 0.022$. Notice that summary(model1) would give us $\hat{\sigma}$ as the (residual standard error).

Hypothesis testing/model reduction. From the anova(model1) output above we see right away that the interaction between vet and tim is not significant ($p = 0.84$). Alternatively we could have fitted the additive model and used anova to compare the two models:

```
> model2 = lm(matter ~ vet + tim,data=organic)
> anova(model2,model1)
Analysis of Variance Table

Model 1: matter ~ vet + tim
Model 2: matter ~ vet + tim + vet:tim
  Res.Df     RSS Df Sum of Sq      F Pr(>F)
1     32 0.66441
2     30 0.65674  2   0.00767 0.1751 0.8402
```

Of course, we get the same test! Note the slightly annoying fact that R uses the notation "Model 1" for the model under the hypothesis and "Model 2" for the full model, no matter what names we have used.

In the same way, we test for the main effect of tim by fitting the model with vet as the only factor and test it against the additive model. Similarly we test for the main effect of vet.

7.1 Analysis of variance (ANOVA)

```
> model3 = lm(matter~vet, data=organic)
> anova(model3,model2)
Analysis of Variance Table

Model 1: matter ~ vet
Model 2: matter ~ vet + tim
  Res.Df    RSS Df Sum of Sq      F    Pr(>F)
1     34 1.43250
2     32 0.66441  2   0.76809 18.497 4.586e-06 ***

> model4 = lm(matter~tim, data=organic)
> anova(model4,model2)
Analysis of Variance Table

Model 1: matter ~ tim
Model 2: matter ~ vet + tim
  Res.Df    RSS Df Sum of Sq      F    Pr(>F)
1     33 2.39596
2     32 0.66441  1   1.73155 83.397 1.991e-10 ***
```

In conclusion, both main effects are highly significant, so the additive model, model2, cannot be reduced any further and is thus the final model.

Parallel tests. The tests above for main effects of vet and tim were both tested on the basis of the additive model, model2. In this sense they were tested in parallel rather than by successive reductions (where the model is updated after each non-significant hypothesis test). All possible parallel tests based on a model, in this case two tests based on model2, can be done with a single call of the function drop1:

```
> drop1(model2, test="F")
Single term deletions

Model:
matter ~ vet + tim
       Df Sum of Sq    RSS     AIC F value    Pr(F)
<none>                0.66441 -135.72
vet     1   1.73155 2.39596  -91.55  83.397 1.991e-10 ***
tim     2   0.76809 1.43250 -112.07  18.497 4.586e-06 ***
```

We recognize the two tests from above. Notice that it must be specified, as shown, that we want to see the F-tests.

An advantage of the function drop1 is that if you apply it to a model which includes an interaction, say between vet and tim, then tests for the main effects will *not* be carried out. In this way we avoid taking the tests for main effects from the wrong model.

Estimates and confidence intervals. As in the one-way ANOVA set-up, the estimates from the final model are extracted by summary, and confidence intervals with confint. First, the summary:

```
> summary(model2)
Coefficients:
              Estimate Std. Error t value Pr(>|t|)
(Intercept)    2.58329    0.04803  53.784  < 2e-16 ***
vetivermectin  0.43863    0.04803   9.132 1.99e-10 ***
tim12         -0.19713    0.05883  -3.351  0.00208 **
tim16         -0.35715    0.05883  -6.071 8.83e-07 ***

Residual standard error: 0.1441 on 32 degrees of freedom
Multiple R-squared:  0.79,Adjusted R-squared: 0.7703
F-statistic: 40.13 on 3 and 32 DF,  p-value: 5.912e-11
```

The control group is used as reference group for treatment, and week 8 is used as reference group for time. Hence, the treatment differences and the time differences are immediately read:

$$\hat{\alpha}(\text{Ivermectin}) - \hat{\alpha}(\text{Control}) = 0.439$$

$$\hat{\beta}(12) - \hat{\beta}(8) = -0.197; \quad \hat{\beta}(16) - \hat{\beta}(8) = -0.357.$$

The corresponding 95% confidence intervals are computed with confint:

```
> confint(model2)
                   2.5 %      97.5 %
(Intercept)    2.4854584   2.6811304
vetivermectin  0.3407918   0.5364638
tim12         -0.3169575  -0.0773092
tim16         -0.4769741  -0.2373259
```

Interaction plots. Interaction plots are easily constructed in R with the function interaction.plot. The commands below produce the two graphs in Figure 7.2.

```
> attach(organic)
> interaction.plot(x.factor=tim, trace.factor=vet,
```

7.1 Analysis of variance (ANOVA)

```
+                         response=matter)   # Figure 7.2, left
> interaction.plot(x.factor=vet, trace.factor=tim,
+                         response=matter)   # Figure 7.2, right
```

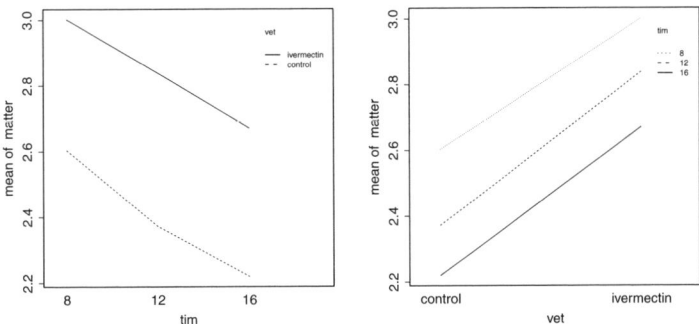

Figure 7.2. Interaction plots for the data on organic matter.

Notice that `interaction.plot` always uses the first argument as a factor. Hence, for a factor with numeric values, R makes the x-axis equidistant (same distance between subsequent levels) even if the values are *not* equidistant. This means that the interaction plot cannot always be used to assess linearity between the factor values used numerically and the response.

The product factor in incomplete designs. Sometimes, not all combinations of the factors occur in designs with two (or more) factors. This causes a little trouble in R. Suppose not all combinations of A and B occur. Nevertheless the product factor A:B is coded in R with all possible levels, and lm gives a "strange" output because it tries to estimate all these although it is obviously not possible form the design. Fortunately, the solution is simple. Write

```
> AB = factor(A:B, exclude=NA)
```

and use this newly constructed product factor instead of A:B in the model. The new factor only contains the combinations that were actually present in the design.

Multi-way ANOVA The analyses are extended to the setting with three or more factors in the natural way. For example, for three factors A, B, and C, the model fits

```
> lm(y ~ A + B + C + A:B + A:C + B:C + A:B:C)
> lm(y ~ A + B + C + A:C)
```

correspond to the model with all interactions, including the three-factor interaction (first command), and a model with all main effects but only one two-factor interaction (second command).

7.2 Regression-type models

We now turn to regression-type models, that is, models with numerical explanatory variables. First we consider the simple linear regression, then quadratic regression and finally models including both numerical variables and factors (categorical variables) as explanatory variables.

7.2.1 Simple linear regression

Suppose we have observed pairs (x_i, y_i) for $i = 1, \ldots, n$ and want to make a linear regression of y on x, $y = \alpha + \beta x + e$. The values below are taken from an example on the relation between digestion of fat (y) and the amount of a particular acid (x). The model is fitted with lm as follows:

```
> x = c(29.8, 30.3, 22.6, 18.7, 14.8, 4.1, 4.4, 2.8, 3.8)
> y = c(67.5, 70.6, 72.0, 78.2, 87, 89.9, 91.2, 93.1, 96.7)
> model1 = lm(y~x)
```

The call is very similar to that of a one-way ANOVA, the only difference is that x is now a numeric variable whereas in the ANOVA setting the explanatory variable was a grouping variable (a factor).

As usual anova gives the analysis of variance table, and summary gives the estimates:

```
> anova(model1)
Analysis of Variance Table

Response: y
          Df Sum Sq Mean Sq F value    Pr(>F)
x          1 896.76  896.76  101.63 2.028e-05 ***
Residuals  7  61.76    8.82

> summary(model1)
```

7.2 Regression-type models

```
Coefficients:
            Estimate Std. Error t value Pr(>|t|)
(Intercept) 96.53336    1.67518   57.63 1.24e-10 ***
x           -0.93374    0.09262  -10.08 2.03e-05 ***

Residual standard error: 2.97 on 7 degrees of freedom
Multiple R-Squared: 0.9356,     Adjusted R-squared: 0.9264
F-statistic: 101.6 on 1 and 7 DF,  p-value: 2.028e-05
```

We read the parameter estimates: $\hat{\alpha} = 96.5$, $\hat{\beta} = -0.93$, $\hat{\sigma} = 2.97$ or $\hat{\sigma}^2 = \text{MSe} = 8.82$. The test of the hypothesis $\beta = 0$ is carried out as a t-test by summary and as a F-test by anova (the tests are of course equivalent as $F = t^2$). The F-test could also have been carried out by fitting the model *without* x and comparing it to model1:

```
> model2 = lm(y~1)
> anova(model2, model1)
Analysis of Variance Table

Model 1: y ~ 1
Model 2: y ~ x
  Res.Df    RSS Df Sum of Sq      F    Pr(>F)
1      8 958.53
2      7  61.76  1    896.76 101.63 2.028e-05 ***
```

7.2.2 Multiple linear regression

In multiple linear regression there are more than one numerical explanatory variables. These are simply added on the right hand side of the "tilde" in the lm call. With response y and regressors x1, x2, and x3, say, we simply write

```
> lm(y ~ x1 + x2 + x3)
```

in order to fit the model given by $y = \beta_0 + \beta_1 x_1 + \beta_2 x_2 + \beta_3 x_3 + e_i$.

Quadratic regression, test for linearity As a special case we may consider *quadratic regression* where a variable and the same variable squared are used as regressors at the same time. We consider data on height and diameter for 18 Corsican Pine trees.[2] It is of interest to

---

[2] The data come from J.N.R Jeffers (1959), *Experimental Design and Analysis in Forest Research*, Almqvist & Wiksell, Stockholm.

describe the height as a function of the diameter as the latter is easier to measure. First, we type the data, make a new variable with the quadratic diameters, and plot height against diameter:

```
> h = c(32,31,30,29,29,28,25,23,
+        20,18,17,17,16,16,15,13,11,11)
> d = c(22.7,22.7,22.6,22.6,21.9,21.9,21.8,21.0,20.4,18.6,
+        19.2,18.9,18.5,18.1,17.7,17.2,16.5,15.5)
> d2 = d^2
> plot(d,h)
```

Interest is in predicting height from diameter, so we use height as response. The quadratic regression and the linear regression are fitted, and the linear model is tested against the quadratic as a test for linearity:

```
> model1 = lm(h ~ d + d2)
> model2 = lm(h ~ d)
> anova(model2, model1)
Analysis of Variance Table

Model 1: h ~ d
Model 2: h ~ d + d2
  Res.Df    RSS Df Sum of Sq      F    Pr(>F)
1     16 4.4535
2     15 1.8640  1    2.5895 20.839 0.000372 ***
```

We conclude that the relationship is not linear and extract the estimates from the quadratic regression model:

```
> summary(model1)
Coefficients:
             Estimate Std. Error t value Pr(>|t|)
(Intercept)  8.727814   1.001295   8.717 2.95e-07 ***
d            0.773303   0.100342   7.707 1.36e-06 ***
d2          -0.010489   0.002298  -4.565 0.000372 ***

Residual standard error: 0.3525 on 15 degrees of freedom
Multiple R-Squared: 0.9803,     Adjusted R-squared: 0.9777
F-statistic: 373.7 on 2 and 15 DF,  p-value: 1.599e-13
```

Hence, with
$$h = \beta_0 + \beta_1 d + \beta_2 d^2 + e,$$
we see that $\hat{\beta}_0 = 8.73$, $\hat{\beta}_1 = 0.773$, $\hat{\beta}_2 = -0.0105$.

7.2 Regression-type models

7.2.3 Test for linearity

In this section we will consider data on hydrolysis of amino acids from the package `Guide1data`, see Appendix C. An experiment was carried out in order to examine the influence of hydrolysis time on the analysis of amino acids. Analyses with 5 types of feed and 5 hydrolysis times were performed. For every combination of feed and hydrolysis time the results are reported for the amino acid serine (in g/16gN). The dataset consists of three variables: the response `serine` and two explanatory variables, `feed` and `hour`.

```
> data(hydrolysis)
> attach(hydrolysis)
```

The `feed` variable is string valued and is thus automatically used as a factor by R. The variable `hour` takes numeric values, and can (and will) be used both as a factor and as a numerical variable. We keep the `hour`-variable as numeric and make a new variable, `hourfac`, as a factor. Moreover, a log-transformation turns out to be appropriate, so we construct a new variable `logserine` with the log 10-transformed values:

```
> hourfac = factor(hour)
> logserine = log10(serine)
```

Model fit. The time variable may be used either as a factor (`hourfac`) or as a numerical variable (`hour`). First, the factor models with and without interaction between `hourfac` and `feed`:

```
> model1 = lm(logserine ~ feed:hourfac)
> model2 = lm(logserine ~ feed + hourfac)
```

The test for interaction is carried out by `anova` as usual:

```
> anova(model2, model1)
Analysis of Variance Table

Model 1: logserine ~ feed + hourfac
Model 2: logserine ~ feed:hourfac
  Res.Df        RSS Df   Sum of Sq      F Pr(>F)
1     41 0.00076722
2     25 0.00050543 16  0.00026179 0.8093 0.6643
```

Next, we turn to models with the time variable as a numerical variable: model3 is the model with different intercepts but a common slope for all feeds (parallel lines), whereas model4 is the model with different intercepts and different slopes:

```
> model3 = lm(logserine ~ feed + hour)
> model4 = lm(logserine ~ feed + hour + feed*hour)
```

Test for linearity. Any two linear models where one is a submodel of the other can be compared by anova. Hence, for example, model3 may be tested against model2. This is a test of the hypothesis that the time-logserine relation is linear.

```
> anova(model3, model2)
Analysis of Variance Table

Model 1: logserine ~ feed + hour
Model 2: logserine ~ feed + hourfac
  Res.Df       RSS Df   Sum of Sq      F Pr(>F)
1     44 0.00080247
2     41 0.00076722  3  0.00003525 0.6279 0.6012
```

Had the interaction between feed and hour been significant, such that model2 had been rejected, a test for linearity could have been carried out by testing model4 against model1. Tests of quadratic models, say, are carried out in the same way: fit the model with the quadratic term(s) and test it against the relevant factor model.

Estimation. The model with parallel lines, model3, turns out to be the final model. If we want the estimates, we may prefer to fit the model without intercept. Estimates are extracted with summary, confidence intervals could be extracted with confint (not shown):

```
> model3a = lm(logserine ~ feed + hour - 1)
> summary(model3a)
Coefficients:
            Estimate Std. Error t value Pr(>|t|)
feedbarley 0.6668516  0.0015819  421.54   <2e-16 ***
feedfish   0.6380504  0.0015819  403.33   <2e-16 ***
feedmais   0.7354692  0.0015819  464.92   <2e-16 ***
feedmeat   0.6588011  0.0015819  416.45   <2e-16 ***
feedsoy    0.7664038  0.0015819  484.47   <2e-16 ***
```

7.3 Model validation

```
hour        -0.0017684  0.0000271  -65.25   <2e-16 ***

Residual standard error: 0.004271 on 44 degrees of freedom
Multiple R-Squared:     1,      Adjusted R-squared:     1
F-statistic: 1.886e+05 on 6 and 44 DF,  p-value: < 2.2e-16
```

Confidence intervals are computed with `confint` as usual. In Section 7.4 we show how to make pairwise comparisons of the feed types with the `estimable` function.

7.3 Model validation

In this section we will use the hydrolysis data again, see Section 7.2.3. Most often model validation is carried out for the initial model which in this case was two-way ANOVA with interaction between `feed` and `hourfac`, i.e., `model1`. One may also prefer to check the final model — this is mostly a matter of taste — and this is what we will do now. The final model was the linear regression model with the time variable `hour` as numerical variable, and with different intercepts but the same slope for all levels of the feed factor, `feed`. The model was fitted with

```
> model3 = lm(logserine ~ feed + hour)
```

7.3.1 Analysis of residuals

The ingredients for the residual analysis can easily be extracted from `model3`: Predicted values are extracted with `predict` (or `fitted`), raw residuals with `residuals (or resid)`, and the standardized residuals with `rstandard`.

```
> pred3 = predict(model3)
> res3 = residuals(model3)
> sres3 = rstandard(model3)
```

The objects may then be used for residual plots (use `plot`) and QQ-plots (use `qqnorm`). The `qqnorm` call below plots the quantiles of `sres` against those of the standard normal distribution. The line with zero intercept and slope one makes the comparison easier. For the residual plots, one may add a horizontal line at zero level with `abline(h=0)`.

```
> plot(pred3, res3)
> plot(pred3, sres3)
> qqnorm(sres3)
> abline(a=0,b=1)
```

The plots are shown in Figure 7.3, and look quite okay.

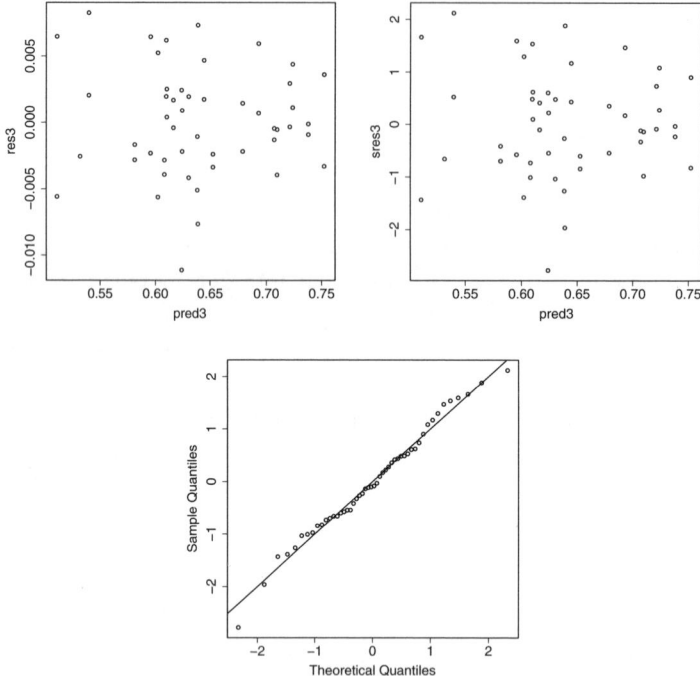

Figure 7.3. Model validation for the hydrolysis data.

7.3.2 Transformation, Box-Cox analysis

Transformation. If the model diagnostics reveal problems with the model assumptions, then a transformation of the response and/or the covariates may sometimes help remedy these. As an example, consider the hydrolysis data once again and remember that the original response, serine, was log-transformed. Consider for a moment the untransformed response, with the same model as above, and construct the residual plot:

```
> model3.orig = lm(serine ~ feed + hour)
> plot(predict(model3.orig), rstandard(model3.orig))
```

7.3 Model validation

The left part of Figure 7.4 shows the plot. There is a clear pattern of large positive residuals for small and large predicted values and many negative residuals for medium predicted values, so the model clearly does not catch the variation in the data. Compare with the upper right graph in Figure 7.3 where there is no such pattern. Hence, the log-transformation seems appropriate for these data.

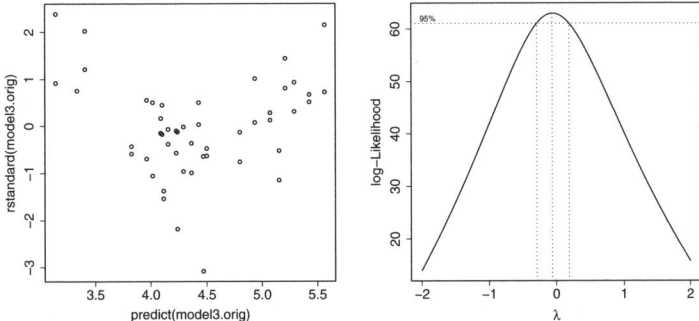

Figure 7.4. Residual plot and Box-Cox plot for the hydrolysis data (untransformed data).

Box-Cox analysis. A Box-Cox analysis can be useful in order to choose a transformation of the data. It compares, in a certain way, power transformations (y^λ) and the logarithmic transformation (corresponding to the power $\lambda = 0$), and chooses "the best" of all those. A Box-Cox analysis is easily carried out in R with the boxcox function. This function is not part of the base package of R, but is part of the add-on package MASS. Hence, this package should be loaded before boxcox can be used; see Appendix B for details on how to use add-on packages.

As argument, boxcox takes a model formula, just like lm. The output is a graph giving the optimal power where zero corresponds to the log-transformation. For the hydrolysis data we write as follows and get the right part of Figure 7.4:

```
> library(MASS)
> boxcox(serine ~ feed + hour)
```

Indeed the optimal value is very close to zero, suggesting a log-transformation (at least for this particular model).

7.4 Estimation of contrasts

We are often interested in estimating certain functions of the parameters of a model. In particular we may be interested in linear functions, also called contrasts. We easily get the estimates themselves "by hand" from the parameter estimates by simply applying the function to the estimates. Usually we want standard errors and/or confidence limits as well for the contrasts, which are not so easy to compute. We want R to help us!

Consider the hydrolysis data again, and assume that we are interested in the difference in log-serine amount between feed types "fish meal" and "maize", that is, we want an estimate of $\alpha(\texttt{mais}) - \alpha(\texttt{fish})$. From the summary output in the end of Section 7.2 we easily get

$$\hat{\alpha}(\texttt{mais}) - \hat{\alpha}(\texttt{fish}) = 0.735 - 0.638 = 0.097$$

but how about a standard error, a confidence interval or a test for the hypothesis of no difference?

For simple contrasts as this one we easily reparameterize (use one of the groups as reference) and use `relevel` as explained in Section 7.1.1. For more complicated contrasts as adjusted means, the `estimable` function can do the job.

Change of reference group. Recall from page 72 that we can change the reference group to `fish` and fit the model once again with the new parameterization. Remember that the model is unchanged, only the parameterization is changed. We get the following:

```
> newfeed = relevel(factor(feed), ref="fish")
> model3.contr = lm(logserine ~ newfeed + hour)
> summary(model3.contr)
Coefficients:
               Estimate Std. Error  t value Pr(>|t|)
(Intercept)    0.6380504  0.0015819  403.33  < 2e-16 ***
newfeedbarley  0.0288013  0.0019099   15.08  < 2e-16 ***
newfeedmais    0.0974188  0.0019099   51.01  < 2e-16 ***
newfeedmeat    0.0207508  0.0019099   10.87 4.86e-14 ***
newfeedsoy     0.1283535  0.0019099   67.21  < 2e-16 ***
hour          -0.0017684  0.0000271  -65.25  < 2e-16 ***
> confint(model3.contr)
                   2.5 %       97.5 %
(Intercept)    0.634862176  0.641238563
newfeedbarley  0.024952192  0.032650345
```

7.4 Estimation of contrasts

```
newfeedmais    0.093569720  0.101267872
newfeedmeat    0.016901682  0.024599835
newfeedsoy     0.124504407  0.132202560
hour          -0.001823011 -0.001713775
```

We find the expected estimate (0.097) as well as a t-test for the hypothesis of no difference ($t = 51.0$, $p = 0$) and a 95% confidence interval $(0.094, 0.101)$.

The estimable function. The `estimable` function is not part of the standard part of R, but is part of the add-on package `gmodels`. Hence, this package should be loaded before `estimable` can be used, see Appendix B for details on how to use add-on packages.

One parameterization of the model is given in `model3a`:

```
> model3a = lm(logserine ~ feed + hour - 1)
> summary(model3a)
Coefficients:
            Estimate Std. Error t value Pr(>|t|)
feedbarley  0.6668516  0.0015819  421.54  <2e-16 ***
feedfish    0.6380504  0.0015819  403.33  <2e-16 ***
feedmais    0.7354692  0.0015819  464.92  <2e-16 ***
feedmeat    0.6588011  0.0015819  416.45  <2e-16 ***
feedsoy     0.7664038  0.0015819  484.47  <2e-16 ***
hour       -0.0017684  0.0000271  -65.25  <2e-16 ***
```

The model is written as $y = \alpha(\text{feed}) + \beta \cdot \text{hour} + e$ and the estimates are given in the above list. We are interested in $\hat{\alpha}(\text{mais}) - \hat{\alpha}(\text{fish})$, which we may also write as

$$0 \cdot \hat{\alpha}(\text{barley}) - 1 \cdot \hat{\alpha}(\text{fish}) + 1 \cdot \hat{\alpha}(\text{fish}) + 0 \cdot \hat{\alpha}(\text{meat}) + 0 \cdot \hat{\alpha}(\text{soy}) + 0 \cdot \hat{\beta}$$

We need to tell R the coefficients (the 1, the -1 and the zeros) of this linear combination.

Below, `contr` defines the proper linear combination and associates a name, `fish-mais` to it. The order of the parameters is the same as in the above `summary` output. More precisely, the `rbind` command makes a row-matrix (a matrix with just one row) and assigns the name `fish-mais` to this row. The `contr` object can then be used as argument to the `estimable` function.

```
> library(gmodels)
```

```
> contr = rbind('fish-meat' = c(0,-1,1,0,0,0))
> estimable(model3a, contr, conf.int=0.95)
          Estimate Std. Error  Pr(>|t|)  Lower.CI Upper.CI
fish-meat 0.0974188    0.00190         0    0.0935    0.101
```

Of course we get the same as with the `relevel` method.

Suppose now that we want to estimate the expected serine amount corresponding to a hydrolysis time of 16 hours, for the feed types "barley" and "meat and bone meal". The expected values could be computed by hand or extracted with `predict`. The corresponding standard errors and confidence intervals may be computed with `estimable` as follows:

```
> barley.est = c(1,0,0,0,0,16)
> meat.est = c(0,0,0,1,1,16)
> est = rbind(barley.est, meat.est)
> estimable(model3a, est, conf.int=0.95)
            Estimate  Std. Error Pr(>|t|)  Lower.CI  Upper.CI
barley.est 0.6385574 0.001405733        0 0.6357243 0.6413904
meat.est   1.3969107 0.002263106        0 1.3923497 1.4014717
```

In this case the `rbind` command (`rbind` for "row-bind") constructs a 2×6-matrix (one row per parameter function) with the coefficients, and a name is associated with each row (each parameter function). The model uses the log-transformed variable as response, so the estimates and confidence limits should now be "back-transformed" (with 10^x) in order to get values on the original scale.

Adjusted means with `estimable`. Adjusted means, also called least squares means or just LS-means, are special linear functions of the parameters and may thus be computed with `estimable`. Consider the hydrolysis data again and let us calculate the adjusted means for the different feeds based on `model3`:

$$y = \mu + \alpha(\text{feed}) + \beta \cdot \text{time} + e.$$

For barley the adjusted mean is given by

$$\hat{\mu} + \hat{\beta} \cdot \overline{\text{time}},$$

and is thus the expected value for the average hydrolysis time in the experiment. For fish it is given by

$$\hat{\mu} + \hat{\alpha}(\text{fish}) + \hat{\beta} \cdot \overline{\text{time}}$$

7.4 Estimation of contrasts

and similarly for the remaining feeds. The adjusted means can be obtained as follows:

```
> mean(c(8,16,24,32,72))
[1] 30.4
> adj.barley = c(1,0,0,0,0,30.4)
> adj.fish = c(1,1,0,0,0,30.4)
> adj = rbind(adj.barley,adj.fish)
> estimable(model3, adj, conf.int=0.95)
            Estimate  Std. Error Pr(>|t|)  Lower.CI  Upper.CI
adj.barley 0.6130925 0.001350477        0 0.6103708 0.6158142
adj.fish   0.5842912 0.001350477        0 0.5815695 0.5870129
```

If we were to calculate the adjusted means for the different feeds based on model2 (the additive model with factors feed and hourfac),

$$y = \mu + \alpha(\texttt{feed}) + \beta(\texttt{time}) + e$$

then, for barley, the adjusted mean is given by

$$\hat{\mu} + \overline{\hat{\beta}(\texttt{time})}.$$

For fish it is given by

$$\hat{\mu} + \hat{\alpha}(\texttt{fish}) + \overline{\hat{\beta}(\texttt{time})},$$

and similarly for the remaining feeds. They are obtained as follows

```
> adj.barley = c(1,0,0,0,0,1/5,1/5,1/5,1/5)
> adj.fish = c(1,1,0,0,0,1/5,1/5,1/5,1/5)
> adj = rbind(adj.barley, adj.fish)
> estimable(model2, adj, conf.int=0.95)
            Estimate  Std. Error Pr(>|t|)  Lower.CI  Upper.CI
adj.barley 0.6130925 0.001367943        0 0.6103299 0.6158551
adj.fish   0.5842912 0.001367943        0 0.5815286 0.5870538
```

In this example we actually get the same adjusted means when using model2 or model3, but the standard errors are slightly different.

8
Models with random effects

In this chapter we will show how to use R for analysis of Gaussian models with random effects and a linear fixed part, the so-called mixed linear models. We will mainly use the `lme` function which works for unbalanced as well as balanced data, but also give some comments on analysis with another functions, `lmer`. This is a newer variant of `lme` where all the facilities from `lme` are not yet built in. Except for the random part, `lme` works quite similarly to `lm`. The fixed part is specified in the same way, estimates are extracted with `summary` and tests can be performed with `anova`. However, confidence intervals are computed with another function, though, namely `intervals`. Before we get started with the analyses, we give some general comments on hypothesis testing in models with random effects.

8.1 F-tests and likelihood ratio tests

Tests of hypotheses in mixed linear models may be carried out as F-tests if the design is balanced in a certain sense. In this case the F-statistic is computed as the ratio of two mean square errors (MSE's). Many experiments are not balanced, though, due to the design itself or due to missing values. Then the exact distributions of the test statistics are no longer F-distributions, and we need approximate methods. Several such methods exist and the most common ones are implemented in R, but some seemingly better methods are not.

Instead, we will carry out likelihood ratio (LR) tests. The rationale for the LR test is the following: For a given model, the maximum of the likelihood function measures (in a certain sense) how well the model fits the data. Hence, if we compare the maximum of the likelihood function under a given model with the maximum of the likelihood function under a null model (assuming a hypothesis to be true), we have a measure of the discrepancy of the models. In other words, we measure how much worse the null model fits the data compared to the full model. This should be compared to the difference in dimensions of the two models: how many more parameters are used in the full models compared to the null model?

To be precise, we use

$$LR = 2 \cdot \log L(\text{full model}) - 2 \cdot \log L(\text{null model})$$

as a test statistic. This statistic is approximately chi-square distributed; the degrees of freedom is equal to the decrement in model dimensions (number of parameters in the model) from the full model to the null model. The chi-square approximation to the distribution of LR is good for large datasets. For small and moderately-sized datasets, however, the experience is that these approximate p-values tend to be too small, thereby sometimes overestimating the importance of certain effects. Therefore it is sometimes recommended to compute a better approximation to the p-value by so-called parametric bootstrap (or simulation) if the approximate p-value is below the significance level, but not very small. This is, as we shall see, quite easy to do with `lme`. Some experience with this will indicate how large a dataset should be for the chi-square approximation to be reasonable.

Some comments on estimation methods: In order to make likelihood ratio tests *for fixed effects* it is absolutely essential that the models are fitted with the maximum likelihood (ML) method. However, it is well-known that the Restricted Maximum Likehood (REML) method generally produced better estimates. Hence, we recommend to always *take the estimates for the final model from the REML fit rather than the ML fit*; also if the model reduction has been carried out by likelihood ratio test using ML. To complicate matters, it is usually recommended to use REML estimation for testing hypotheses about the random part of the model, i.e., use

$$REML.LR = 2 \cdot \log ReL(\text{full model}) - 2 \cdot \log ReL(\text{null model})$$

where $\log ReL$ stands for the restricted log-likelihood function. Notice that `R` as default uses REML estimation.

8.2 Analysis of linear mixed models (`lme`, `lmer`)

8.2.1 A single random factor

In this section we will use the dataset `porkers` on tenderness of pork chops from `Guide1data`, see Appendix C. Suppose that the four relevant variables are available: the response variable `tenderness` as well as the factors `porker` with levels $1, 2, \ldots, 24$, `ph` with levels `low` and `high`, and `chilling` with levels `fast` and `tunnels`.

Model fit. First the package `nlme` must be loaded so that we can use `lme`, see Appendix B. Second, we use `lme` to fit model with interaction between `ph` and `chilling` and with `porker` as a random factor with random effects. The fixed effect part of a model is written in the usual `lm` way. The random part is specified by a one-sided expression followed by some grouping variables after the `|`. Here, we have only one grouping variable, `porker`. For a start, we use REML estimation (which is default, so the option `method="REML"` could be skipped).

```
> data(porkers)
> library(nlme)
> model1 = lme(tenderness ~ chilling + ph + ph:chilling,
+              random =~ 1|porker, method="REML",
+              data=porkers)
```

Model validation. The model fitted in `model1` has the form $y_i = \mu_i + \eta(\text{porker}_i) + e_i$ where the mean μ_i is given by the interaction between chilling method and pH value, $\eta(1), \ldots, \eta(24)$ are independent random effects, and e_i are independent error terms. Of course, `model1` contains estimates of the fixed effects and thus of the μ_i's, but `lme` also computes predictions of the random effects, the so-called BLUPS denoted $\hat{\eta}(1), \ldots, \hat{\eta}(24)$.

The residuals $y_i - \hat{\mu}_i - \hat{\eta}(\text{porker}_i)$ can thus be interpreted as a prediction of the error term e_i, and we can check the assumptions on variance homogeneity and normality on these residuals. The commands below produce the plots in Figure 8.1. Notice that R uses a standardized version of the residuals. It may seem like the variation is smaller for large fitted values, but we will not pursue this any further.

```
> plot(model1)
```

```
> qqnorm(model1)
```

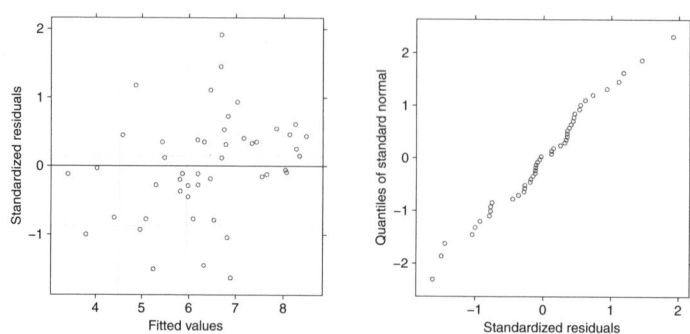

Figure 8.1. Residual plot and QQ-plot for model model1.

Estimation, confidence intervals. The summary function is used in the usual way to extract the estimates for the parameters from the fixed part of the model as well as for the variance parameters from the random part:

```
> summary(model1)
Linear mixed-effects model fit by REML

Random effects:
 Formula: ~1 | porker
         (Intercept)  Residual
StdDev:    1.118207   0.6813447

Fixed effects: tenderness ~ chilling + ph + ph:chilling
                        Value Std.Error DF  t-value p-value
(Intercept)           7.010000 0.3780010 22 18.544922  0.0000
chillingtunnel        0.212500 0.2781578 22  0.763955  0.4530
phlow                -1.545833 0.5345742 22 -2.891710  0.0085
chillingtunnel:phlow  0.165000 0.3933745 22  0.419448  0.6790
```

The fixed part parameters are interpreted in the same way as for lm fits. Moreover, the standard deviation for the random porker factor is estimated to 1.12, and the residual standard deviation is estimated to 0.68.

Finally, intervals extracts the confidence intervals for all parameters (fixed and random):

8.2 Analysis of linear mixed models (lme, lmer)

```
> intervals(model1)
Approximate 95% confidence intervals

 Fixed effects:
                          lower       est.      upper
 (Intercept)           6.2260738   7.010000  7.7939262
 chillingtunnel       -0.3643639   0.212500  0.7893639
 phlow                -2.6544724  -1.545833 -0.4371943
 chillingtunnel:phlow -0.6508088   0.165000  0.9808088

 Random Effects:
  Level: porker
                     lower      est.     upper
 sd((Intercept)) 0.7840528  1.118207  1.594774

 Within-group standard error:
     lower       est.      upper
 0.5070156  0.6813447  0.9156139
```

Hypothesis tests. We use anova to one model against another (nested models). Remember to fit the model with ML rather than REML, see Section 8.1. Here we test the additive model (no interaction ph and chilling) against the full model:

```
> model1.ML = lme(tenderness ~ chilling+ph+ph:chilling,
+                 random =~ 1|porker, method="ML",
+                 data=porkers)
> model2.ML = lme(tenderness ~ chilling + ph,
+                 random =~ 1|porker, method="ML",
+                 data=porkers)
> anova(model1.ML, model2.ML)
          Model df    AIC      BIC    logLik   L.Ratio  p-value
model1.ML     1  6 151.709 162.936 -69.8548
model2.ML     2  5 149.900 159.256 -69.9504 0.1911661   0.6619
```

From the output we see that the maximum of the log-likelihood function is -69.85 in the model with interaction (model1.LM) and -69.95 in the additive model (model2.LM). This gives $LR = 0.19$ which, evaluated in the chi-square distribution with one degree of freedom, amounts to a p-value of 0.66. Similarly we test for the effect of chilling ($p = 0.13$) and thereafter for the effect of ph ($p = 0.0048$):

```
> model3.ML = lme(tenderness ~ ph, random =~ 1|porker,
```

```
+                       method="ML", data=porkers)
> anova(model3.ML, model2.ML)
          Model df     AIC     BIC   logLik L.Ratio p-value
model3.ML     1  4 150.219 157.704 -71.1098
model2.ML     2  5 149.900 159.256 -69.9504 2.318832  0.1278

> model4.ML = lme(tenderness ~ 1, random =~ 1|porker,
+                       method="ML", data=porkers)
> anova(model4.ML, model3.ML)
          Model df     AIC     BIC   logLik L.Ratio p-value
model4.ML     1  3 156.190 161.804 -75.0953
model3.ML     2  4 150.219 157.704 -71.1098 7.971096  0.0048
```

Computation of p-value with parametric bootstrap (simulation). The above p-values are either well above or well below 5%, so we do not doubt the conclusions based on the chi-square approximations. For later use let us show how to compute a more accurate approximation to the p-value, anyway.

Take the test of `model3.ML` against `model2.ML`, say, where we got the value 2.319 of LR. First, `simulate.lme` is used to to simulate 1000 datasets from the null model, using the estimates from the real dataset. In our case the null model corresponds to `model3.ML`. For each simulated dataset, the null as well as the alternative model, here given by `model2.ML`, are fitted. R saves the maximum values of the log-likelihood function for each simulated dataset in a list. This takes roughly half a minute (depending on the computer, of course). We plug out the relevant values, compute the LR test statistic in `lrsim`, and finally compute the frequency of simulated LR values that are larger than our observed value, 2.319. In this case out bootstrap p-value is 0.133, slightly larger than the approximate value 0.1278.

```
> sim = simulate.lme(model3.ML, m2=model2.ML, nsim=1000,
+                       method="ML")
> lrsim = 2*(sim$alt$ML - sim$null$ML)
> psim = sum(lrsim > 2.319)/1000
> psim
[1] 0.133
```

Analysis of the final model. We use $y_i = \alpha(\text{ph}_i) + \eta(\text{porker}_i) + e_i$ (`model3.ML`) with $\eta_1, \ldots, \eta_{24}$ Gaussian random effects, as the final model. We fit the model with REML and extract the estimates:

8.2 Analysis of linear mixed models (lme, lmer)

```
> summary(model3)
Linear mixed-effects model fit by REML

Random effects:
 Formula: ~1 | porker
         (Intercept)  Residual
StdDev:    1.116364   0.6873579

Fixed effects: tenderness ~ ph
              Value  Std.Error  DF  t-value  p-value
(Intercept)  7.116250 0.3514849 24 20.24624  0.0000
phlow       -1.463333 0.4970748 22 -2.94389  0.0075
```

As for linear models, we may of course reparameterize the model in order to obtain estimates of other contrasts. Moreover, the `estimable` function works for `lme` objects, too.

8.2.2 Two or more random factors

Two nested random factors. Consider the dataset `vitE` from the package `Guide1data`, see Appendix C, on concentration of vitamin E in meat samples. The dataset contains three variables, `lab`, `sample` and `vitE`. In the example the square root of `vitE` is used as response, so we construct a new variable with those values. Moreover, we construct factors of the original numerical variables, assuming that the data has first been loaded and attached:

```
> sqrtEvit = sqrt(vitE)
> L = factor(lab)
> S = factor(sample)
```

The start model has `S` as fixed factor, and `L` and `L:S` as random factors. `lme` will not accept product factors with `:` in the random part, so the product factor is constructed and called `LS` before the `lme` call. Naturally, `LS` is a subdivision of `L`; this is indicated to R with `L/LS` in the random part of the model specification.

```
> LS = L:S
> library(nlme)
> model1 = lme(sqrtEvit ~ S, random =~ 1|L/LS)
```

It turns out that the effect of `S` is indeed significant (try it!) so the model cannot be reduced. In order to get the estimates in a more direct form we fit the model without intercept:

```
> model1a = lme(sqrtEvit ~ S - 1, random =~ 1|L/LS)
> summary(model1a)
Linear mixed-effects model fit by REML

Random effects:
 Formula: ~1 | L
        (Intercept)
StdDev:   0.2452144

 Formula: ~1 | LS %in% L
        (Intercept)  Residual
StdDev:   0.1347001 0.0677408

Fixed effects: sqrtEvit ~ S - 1
      Value  Std.Error DF  t-value   p-value
S1 1.177684 0.1269398 16  9.277495      0
S2 2.541707 0.1269398 16 20.022923      0
S3 2.131801 0.1269398 16 16.793791      0
S4 1.777835 0.1269398 16 14.005333      0
S5 2.642682 0.1269398 16 20.818383      0
```

In particular we see that the standard deviation associated to L and L:S are 0.245 and 0.135, respectively, whereas the residual standard error is 0.068.

The general case, analysis with lmer. More generally, the random factors may be non-nested and there may be more than two random factors. lme can handle such cases, too, but the syntax is quite complicated. In such cases it is much easier to use the lmer function from the lme4 package. It is, of course, also applicable in the simple situations, but note that there is no such function as simulate.lmer for parametric bootstrap of the p-value and that the estimable function does not seem to work with lmer, either.

Let us consider the data chocolate from the package Guide1data, see Appendix C, on sweetness of chocolate. The dataset contains four variables: product, session, assessor and sweetness score score. First, the explanatory variables are made factors and the score is transformed with arcsin to stabilize variation. Again, we assume that the data has first been loaded and attached.

```
> y = asin(sqrt(score/15))
```

8.2 Analysis of linear mixed models (lme, lmer)

```
> A = factor(assessor)
> P = factor(product)
> S = factor(session)
```

Then consider the model with all main effects and two-factor interactions with P as fixed effect and A, S, A × P, P × S and A × S as random effects. The syntax for the fixed part of the model is the usual. The random factors are specified as (1|random.factor):

```
> library(lme4)
> model1 = lmer(y ~ P + (1|A)+(1|S)+(1|A:P)+(1|P:S)+(1|A:S),
+               method="REML")
Warning message:
Estimated variance for factors 'P:S', 'S' is effectively zero
in: 'LMEoptimize<-'('*tmp*', value = list(maxIter = 200,
        tolerance = 1.49011
611938477e-08,
> summary(model1)
Linear mixed-effects model fit by REML
Formula: y ~ P + (1 | A) + (1 | S) + (1 | A:P) +
              (1 | P:S) + (1 | A:S)
  AIC   BIC logLik MLdeviance REMLdeviance
 92.1 122.9 -36.05      60.05        72.1
Random effects:
 Groups    Name         Variance    Std.Dev.
 A:P       (Intercept)  7.2735e-02  2.6969e-01
 A:S       (Intercept)  7.6564e-03  8.7501e-02
 P:S       (Intercept)  2.2309e-11  4.7232e-06
 A         (Intercept)  6.6916e-02  2.5868e-01
 S         (Intercept)  2.2309e-11  4.7232e-06
 Residual                4.4618e-02  2.1123e-01
number of obs: 160, groups: A:P, 40; A:S, 32; P:S, 20;
        A, 8; S, 4

Fixed effects:
            Estimate Std. Error t value
(Intercept)   0.8533     0.1382   6.176
P2           -0.3469     0.1448  -2.396
P3           -0.3124     0.1448  -2.157
P4            0.1418     0.1448   0.979
P5           -0.3581     0.1448  -2.473
```

R writes a warning that two of the variance components are extremely close to zero. This is also seen from the summary output: the estimates for σ^2_{PS} and σ^2_A are 10^{-11}! Hence, we fit the model without these two

factors (model2 and model2.ML below). We are interested in the effect of
product, so we carry out a likelihood ratio test for the effect. Remember
that we should then use the ML method for estimation.

```
> model2 = lmer(y ~ P + (1|A) + (1|A:P) + (1|A:S))
> model2.ML = lmer(y ~ P + (1|A) + (1|A:P) + (1|A:S),
+                  method="ML")
> model3.ML = lmer(y ~ 1 + (1|A) + (1|A:P) + (1|A:S),
+                  method="ML")
> anova(model3.ML, model2.ML)
Data:
Models:
model3.ML: y ~ 1 + (1 | A) + (1 | A:P) + (1 | A:S)
model2.ML: y ~ P + (1 | A) + (1 | A:P) + (1 | A:S)
          Df    AIC     BIC  logLik  Chisq Chi Df Pr(>Chisq)
model3.ML  4 85.237  97.537 -38.618
model2.ML  8 75.793 100.394 -29.896 17.444      4   0.001584
```

The model without product effect is rejected so model2 is the final model.
The estimates are extracted with summary:

```
> summary(model2)
Linear mixed-effects model fit by REML

Random effects:
 Groups   Name        Variance  Std.Dev.
 A:P      (Intercept) 0.0729008 0.270001
 A:S      (Intercept) 0.0076823 0.087649
 A        (Intercept) 0.0664757 0.257829
 Residual             0.0445939 0.211173
number of obs: 160, groups: A:P, 40; A:S, 32; A, 8

Fixed effects:
            Estimate Std. Error t value
(Intercept)   0.8533     0.1380   6.182
P2           -0.3469     0.1450  -2.393
P3           -0.3124     0.1450  -2.155
P4            0.1418     0.1450   0.979
P5           -0.3581     0.1450  -2.470
```

Note that no p-values are associated with the t-values. This is because,
in general, these statistics do not vary according to a t-distribution. Still,
the t-values give an idea about the significance of the parameters. For
the same reason there is no function like intervals applicable for lmer

8.2 Analysis of linear mixed models (lme, lmer)

objects. Note also that `estimable` is not immediately applicable for `lmer` objects.

If we want estimates of the product mean values directly, we fit the model without an intercept:

```
> model2a = lmer(y ~ P - 1 + (1|A) + (1|A:P) + (1|A:S))
```

9
Repeated measurements

This chapter is about the analysis of repeated measurements, i.e. data where the response has been measured multiple times for each individual. If data are collected over time, as is often the case, such data are often called longitudinal data. Throughout the chapter we will use the dataset goats from the Guide1data package, see Appendix C. The data concern the growth of goats.

9.1 Preliminaries

9.1.1 Wide form and long form

The dataset goats contains repeated weight recordings on 28 goats. The goats receive one out of four treatments (feed with values 1–4), and their weight is recorded prior to treatment (w0), and then again at day 26, 45 and 91 after receiving the treatment (w26, w45, w61, w91). The structure of the data is thus

```
> data(goats)
> head(goats)
  goat feed   w0  w26  w45  w61  w91
1    1    1 20.4 21.0 21.5 21.3 22.3
2    2    1 10.3 11.4 11.6 12.0 12.5
3    3    1 12.5 13.3 14.5 14.5 15.4
4    4    1 10.8 11.4 11.7 12.1 12.5
5    5    1 13.6 15.2 15.0 15.5 15.8
6    6    1 19.0 19.4 19.8 19.4 19.6
```

This is sometimes referred to as the data being on *wide form*, with one line per unit (in this case goat). Most of the statistical analysis we will be doing, however, requires the data to be on so-called long form with one line per recorded weight (post treatment). The long data format can be obtained as follows:

```
> long.g = reshape(goats, idvar="goat", varying=list(4:7),
+                  times=c(26,45,61,91), v.names="weight",
+                  direction="long")
> long.g = long.g[order(long.g$goat),] # sorting the data
> head(long.g)
     goat feed   w0 time weight
1.26    1    1 20.4   26   21.0
1.45    1    1 20.4   45   21.5
1.61    1    1 20.4   61   21.3
1.91    1    1 20.4   91   22.3
2.26    2    1 10.3   26   11.4
2.45    2    1 10.3   45   11.6
```

We see that we now have one data line per observation with the baseline (w0) as a variable, and we have also constructed a new variable, time.

9.1.2 Profile plots

In order to get an overview of the data it is useful to make some plots of the repeated measurements. It is recommended to make *subject profiles* (for each subject or individual) as well as *average profiles* (for each treatment). The subject profiles give an impression of the "typical" time-response relationship whereas the average profiles illustrate, among others, potential interactions between treatment and time.

Subject profiles. For the subject profiles we plot the response against time for each profile, either in the same plot or in one plot per treatment. We may wish to include the baseline weight measurement in the plot. The following program lines produces the left part of Figure 9.1:

```
> long.gb = reshape(goats, idvar="goat", varying=list(3:7),
+                   times=c(0,26,45,61,91), v.names="weight",
+                   direction="long") # Dataset with baseline
+                                     # included as response
> library(lattice)
> xyplot(weight~time|feed, groups=goat, xlab="Time (days)",
+        type="l", data=long.gb)
```

9.2 Analysis of summary measures

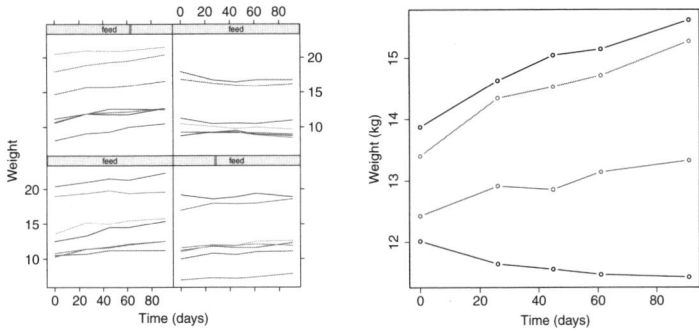

Figure 9.1. Subject profiles (left) and average profiles (right) for the goat data.

Average treatment profiles. The average profiles illustrate treatment differences, both at separate time points and over time. The latter corresponds to interaction between time and treatment. The following commands produce the right part of Figure 9.1:

```
> avedata = with(long.gb,                            # Averages
+               tapply(weight,list(time,feed), mean))
> avedata
          1        2        3        4
0  13.87143 12.42857 13.40000 12.01429
26 14.62857 12.91429 14.34286 11.64286
45 15.04286 12.85714 14.52857 11.55714
61 15.14286 13.14286 14.71429 11.47143
91 15.61429 13.32857 15.27143 11.42857

> obsTimes = c(0,26,45,61,91)
> plot(obsTimes, avedata[,1], ylim=range(avedata),
+      xlab="Time (days)", ylab="Weight (kg)",
+      type="b")                                     # Feed type 1
> for (i in 2:4) lines(obsTimes, avedata[,i],
+                      col=i, type="b")  # Feed types 2-4
```

Notice how the axes are adjusted in the `plot` command such that there is room enough for all data.

9.2 Analysis of summary measures

We talk about summary measures when, for each subject or profile, the collection of observations is reduced to a single or a few numbers. Analysis of the summaries is simpler as they can often be assumed to be independent. There are many possible summary measures. Here we will use the increment in weight from day 26 to day 91 for illustration of the procedure. The analysis is easily carried out with lm using the original data on the wide form:

```
> mIncr = lm(w91-w26 ~ w0+factor(feed), data=goats)
> summary(mIncr)
Coefficients:
                 Estimate Std. Error t value Pr(>|t|)
(Intercept)      1.008993   0.351248   2.873   0.0086 **
w0              -0.001678   0.022110  -0.076   0.9402
factor(feed)2   -0.573850   0.244206  -2.350   0.0277 *
factor(feed)3   -0.057934   0.242338  -0.239   0.8132
factor(feed)4   -1.203117   0.245571  -4.899 5.99e-05 ***

Residual standard error: 0.453 on 23 degrees of freedom
```

Notice that the baseline measurements (w0) are included as a covariate in the model. We see, among other things, that treatment 4 gives a significant smaller weight increase (from day 26 to 91) when compared to treatment 1 ($p<0.001$).

9.3 The random intercepts model

In the rest of the chapter we use all the repeated measurements data and not just parts of them as in the previous section. It seems reasonable to include the subject (here goat) in the model as a random factor. The simplest such model is the random intercepts model which can be fitted with the lme or lmer function, see Section 8.2. For example, the model with interaction between feed type and observation day as factor, and with baseline measurements (w0) as covariate is fitted with

```
> library(nlme)
> mRandInt = lme(weight~w0+factor(feed)*factor(time),
+                random=~1|goat, data=long.g)
```

This model can then be used for analysis in the usual way (see Section 8.2.1 for an example).

9.4 Investigation of the correlation structure

In the random intercepts model, mRandInt, any two observations from the same goat have the same covariance (and thus the same correlation). For repeated measures this is often not a reasonable assumption, and it is recommended to examine the empirical correlation structure. This is done using the residuals from a linear model.

It would be natural to use the model with the the same fixed effects as mRandInt, namely:

```
> lmFit0= lm(weight~w0+factor(feed)*factor(time),data=long.g)
```

Residuals from this model are on the long form, but we need them on the wide form. We could convert them with reshape but it is perhaps easier to use following model fit:

```
> lmFit = lm(data.matrix(goats[,4:7]) ~
+                    goats$w0 + factor(goats$feed))
> lmFit

Coefficients:
                        w26       w45       w61       w91
(Intercept)         1.49847   1.97711   2.20002   2.50746
goats$w0            0.94656   0.94192   0.93306   0.94488
factor(goats$feed)2 -0.34854  -0.82666  -0.65373  -0.92239
factor(goats$feed)3  0.16052  -0.07024   0.01130   0.10259
factor(goats$feed)4 -1.22782  -1.73644  -1.93861  -2.43094
```

Some explanation is needed: The object data.matrix(goats[,4:7]) is a matrix of dimension 28 times 4 containing the measurements from days 26, 45, 61 and 91 in separate columns. When this matrix is used as response in lm, a linear model is fitted to each column at a time, i.e. for each day separately. This means that different coefficients are obtained for the different days. This corresponds to the interaction between factor(feed)*factor(time) in lmFit0. However, different coefficients are allowed for the baseline effect, w0, as well so the fixed effects part in lmFit0 and lmFit are not exactly the same. In this case, the four estimates are almost the same, so it makes very little difference.

The residuals from lmFitRes are extracted with residuals as usual. Notice that they are automatically on the wide form:

```
> lmFitRes = residuals(lmFit)
```

```
> head(lmFitRes)
        w26         w45          w61         w91
1  0.19175994  0.30776433  0.06561108  0.5170017
2  0.15199128 -0.07886451  0.18949066  0.2602836
3 -0.03043535  0.74891603  0.63676441  1.0815489
4 -0.32128750 -0.44982348 -0.17703803 -0.2121561
5  0.82835134  0.21280630  0.61040129  0.4421816
6 -0.08305948 -0.07355056 -0.52810858 -0.8601672
```

The following commands compute the variance-covariance matrix and the corresponding correlation matrix, respectively:

```
> var(lmFitRes)    # Variance-covariance matrix
        w26        w45        w61        w91
w26 0.1569817 0.1286832 0.1124339 0.1478628
w45 0.1286832 0.1749707 0.1355428 0.1620887
w61 0.1124339 0.1355428 0.1632543 0.1780422
w91 0.1478628 0.1620887 0.1780422 0.3135149
> cor(lmFitRes)    # Correlation matrix
        w26        w45        w61        w91
w26 1.0000000 0.7764519 0.7023289 0.6665079
w45 0.7764519 1.0000000 0.8019760 0.6920558
w61 0.7023289 0.8019760 1.0000000 0.7869762
w91 0.6665079 0.6920558 0.7869762 1.0000000
```

From the diagonal of the variance-covariance matrix we see that the variation is roughly the same for days 26, 45, and 61, but somewhat larger for day 91. This is slightly worrying as all the models in Section 9.5 below assume variance homogeneity. In Section 9.6, we fit models that allow for variance inhomogeneity.

The empirical correlation matrix shows that, in general, the estimated correlation is smaller for observations that are far apart in time compared to observations that are close in time. This pattern is more clearly recognized in Figure 9.2 where the correlations are plotted against the time distances.

The figure is produced by the following commands, where the correlations are extracted from the matrix, and the corresponding time distances are constructed:

```
> corMat = cor(lmFitRes)              # Matrix of correlations
> corVec = corMat[upper.tri(corMat)]  # Vector with
>                                     # off-diagonal elements
> corVec
```

9.4 Investigation of the correlation structure

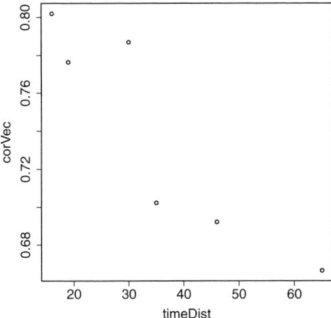

Figure 9.2. Estimated correlations from the linear model lmFit.

```
[1] 0.776451 0.702328 0.801976 0.666507 0.692055 0.786976
> timeDist = c(45,61,61,91,91,91) - c(26,26,45,26,45,61)
> timeDist
[1] 19 35 16 65 46 30
> plot(timeDist, corVec)
```

The pattern in the correlations is also revealed in pairwise scatterplots of the residuals (Figure 9.3), produced by pairs(lmFitRes). Such a plot might also be useful for detecting potential outliers.

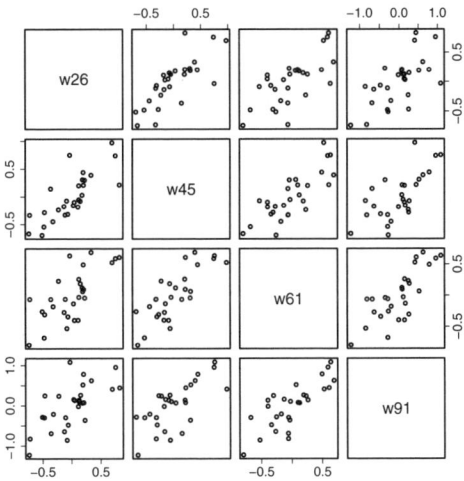

Figure 9.3. Residuals from the linear model lmFit.

9.5 Serial correlation and variance homogeneity

In the random intercepts model, any two observations from the same goat have the same covariance. We saw above that this might be an inappropriate assumption for the goat data, and this is often the case for repeated measures. It would be more appropriate to use a model with serial correlations, such that the covariance decreases with the time distance between the observations.

There are many possibilities for specifications of such models; some of them are given in Table 9.1. The names in the first column correspond to the model objects below. The second and third column specify the variance of an observation and the covariance between two observations with a time distance h apart (from the same subject), respectively. In particular, all model assume variance homogeneity (same variance for all observations). The term ν^2 comes from random effects, τ^2 comes from serially correlated residuals, and σ^2 comes from independent residuals (measurement error). The fourth and fifth columns report whether the model has contributions from random effects and measurement error, respectively, and the type of serial correlation is reported in the last column.

The model mGausNugget with contributions from a random intercept, Gaussian serial correlation and independent errors is sometimes referred to at the Diggle model. Notice that exponential serial correlation is equivalent to an AR(1) process if the time points are equidistant. For completeness we have included the random intercepts model (mRandInt) and the unrestricted model (mUnrestrict) that allows the covariances to vary freely (no pre-specified functional form) in the table.

9.5.1 Fitting models with serial correlation structure

The models containing random intercept of goat are all fitted with lme. Compared to the simple random intercepts model mRandInt above, a corr option must be supplied. For example the Diggle model is fitted with

```
> mGausNugget = lme(weight~w0+factor(feed)*factor(time),
+                   random=~1|goat,
+                   corr=corGaus(form=~time|goat,nugget=TRUE),
+                   data=long.g)
```

9.5 Serial correlation and variance homogeneity

Table 9.1. Models with serial correlation. The covariance in the third column is the assumed covariance between two observations from the same subject with a time difference h. See the main text for details.

Model name	Model specification		Model parts		
	Variance	Covariance	RI	Meas. error	Serial corr.
mGausNugget	$\nu^2 + \tau^2 + \sigma^2$	$\nu^2 + \tau^2 \exp(-(h/\phi)^2)$	Yes	Yes	Gaussian
mGausNoNugget	$\nu^2 + \tau^2$	$\nu^2 + \tau^2 \exp(-(h/\phi)^2)$	Yes	No	Gaussian
mGausNoRI	τ^2	$\tau^2 \exp(-(h/\phi)^2)$	No	No	Gaussian
mExpNugget	$\nu^2 + \tau^2 + \sigma^2$	$\nu^2 + \tau^2 \exp(-h/\phi)$	Yes	Yes	Exponential
mExpNoNugget	$\nu^2 + \tau^2$	$\nu^2 + \tau^2 \exp(-h/\phi)$	Yes	No	Exponential
mExpNoRI	τ^2	$\tau^2 \exp(-h/\phi)$	No	No	Exponential
mRandInt	$\nu^2 + \sigma^2$	ν^2	Yes	Yes	No
mUnrestrict	$\nu^2 + \tau^2$	$\nu^2 + \tau^2 \rho_{ij}$	Yes	Yes	Any

The fixed part and the random intercepts part of the model is specified in the usual way. The `corr` option specifies that correlation is decreasing as an Gaussian density (`corGaus`), that time is measured by the `time` variable, and `nugget=TRUE` ensures that iid. errors are included in the model.

The other models with random effects and serial correlation are fitted similarly. Only the `corr` option should be changed:

```
mGausNoNugget:  corr=corGaus(form=~time|goat, nugget=FALSE)
mExpNugget:     corr=corExp(form=~time|goat, nugget=TRUE)
mExpNoNugget:   corr=corExp(form=~time|goat, nugget=FALSE)
mUnrestrict:    corr=corSymm(form=~1|goat)
```

The models without random effect (but with serial correlation) must be fitted with `gls`, which is also included in the `nlme` package:

```
> mGausNoRI = gls(weight~w0+factor(feed)*factor(time),
+                 corr=corGaus(form=~time|goat), data=long.g)
> mExpNoRI = gls(weight~w0+factor(feed)*factor(time),
+                corr=corExp(form=~time|goat), data=long.g)
```

9.5.2 Extracting the estimates

Estimates for the parameters (both for fixed and random effects) are obtained by `summary`. In this case `summary` gives a lot of output; we only show what is relevant at the moment:

```
> summary(mGausNugget)
Linear mixed-effects model fit by REML

Random effects:
 Formula: ~1 | goat
         (Intercept)  Residual
StdDev:    0.399672  0.3026485

Correlation Structure: Gaussian spatial correlation
 Formula: ~time | goat
 Parameter estimate(s):
     range       nugget
 36.3141567   0.3420622

Fixed effects: weight ~ w0 + factor(feed) * factor(time)
```

9.5 Serial correlation and variance homogeneity

```
                                  Value Std.Error DF   t-val    p-val
(Intercept)                       1.546 0.357     72   4.324    0.0000
w0                                0.943 0.021     23  43.112    0.0000
factor(feed)2                    -0.353 0.269     23  -1.310    0.2030
factor(feed)3                     0.158 0.268     23   0.592    0.5593
factor(feed)4                    -1.234 0.271     23  -4.554    0.0001
factor(time)45                    0.414 0.114     72   3.623    0.0005
factor(time)61                    0.514 0.139     72   3.695    0.0004
factor(time)91                    0.985 0.159     72   6.176    0.0000
factor(feed)2:factor(time)45     -0.471 0.161     72  -2.915    0.0047
factor(feed)3:factor(time)45     -0.228 0.161     72  -1.413    0.1618
factor(feed)4:factor(time)45     -0.500 0.161     72  -3.091    0.0028
factor(feed)2:factor(time)61     -0.285 0.196     72  -1.451    0.1509
factor(feed)3:factor(time)61     -0.142 0.196     72  -0.725    0.4703
factor(feed)4:factor(time)61     -0.685 0.196     72  -3.483    0.0008
factor(feed)2:factor(time)91     -0.571 0.225     72  -2.531    0.0135
factor(feed)3:factor(time)91     -0.057 0.225     72  -0.253    0.8009
factor(feed)4:factor(time)91     -1.200 0.225     72  -5.316    0.0000
```

The interpretation of the fixed effects parameters is the usual one, but some explanation for the random effects parameters is needed. Recall from Table 9.1 that the variance of a measurement is $\nu^2 + \sigma^2 + \tau^2$, and that the correlation between two measurements with time distance h is $\nu^2 + \tau^2 \exp(-(h/\phi)^2)$.

The intercept standard deviation (0.3997) is the estimate of ν. The square of the residual standard deviation (0.3026^2) is the estimate of the sum $\sigma^2 + \tau^2$. The range parameter estimate (36.31) is the estimate of ϕ. The nugget parameter estimate (0.3421) is the estimate of $\sigma^2/(\sigma^2 + \tau^2)$, that is, the fraction of the residual variance that is due to independent measurement noise. Hence, we have the equations:

$$\hat{\nu}^2 = 0.3997^2 = 0.1598, \quad \hat{\sigma}^2 + \hat{\tau}^2 = 0.3026^2 = 0.0916,$$

$$\hat{\phi} = 36.31, \quad \frac{\hat{\sigma}^2}{\hat{\tau}^2 + \hat{\sigma}^2} = 0.3421.$$

Solving for the original parameters, we find

$$\hat{\nu}^2 = 0.1598, \quad \hat{\tau}^2 = 0.0603, \quad \hat{\phi} = 36.31, \quad \hat{\sigma}^2 = 0.0313.$$

These values determine the correlation as a function of the time distance. In particular, we may compute the correlations or the covariances for the observed time distances (here 16, 19, 30, 35, 46, 65). The variance-covariance matrix can also be extracted with getVarCov:

```
> getVarCov(mGausNugget, individuals=1, type="marginal")
goat 1
Marginal variance covariance matrix
         1       2       3       4
1  0.25133 0.20557 0.18354 0.16218
2  0.20557 0.25133 0.20937 0.17185
3  0.18354 0.20937 0.25133 0.19019
4  0.16218 0.17185 0.19019 0.25133
  Standard Deviations: 0.50133 0.50133 0.50133 0.50133
```

The argument `individuals=1` gives the value of a single individual, which individual does not matter here. The value 1 is the default, but if no individual is labeled 1, it is necessary to provide another value.

9.5.3 Hypothesis tests in fixed part of the model

Hypotheses about the fixed effects are tested as usual: First the "full" model and the model under the hypothesis are both fitted with ML; then they are compared with `anova`.

Consider for example the hypothesis that the time-response relationship is linear. First the model with baseline measurements and the interaction between treatment (feed type) and the time factor is fitted, this time with ML. Then a corresponding model, but now with time as a numerical variable, is fitted. This corresponds to a linear relationship between time and weight with treatment-dependent intercepts and slopes.

```
> mGausNuggetML = lme(weight~w0+factor(feed)*factor(time),
+        random=~1|goat,
+        corr=corGaus(form=~time|goat,nugget=TRUE),
+        data=long.g, method="ML")

> mGausNuggetLinML = lme(weight~w0+factor(feed)*time,
+        random=~1|goat,
+        corr=corGaus(form=~time|goat,nugget=TRUE),
+        data=long.g, method="ML")

> anova(mGausNuggetLinML, mGausNuggetML)
                 Model df    AIC     BIC  logLik L.Rat p-val
mGausNuggetLinML     1 13 89.7161 125.056 -31.85
mGausNuggetML        2 21 96.8577 153.946 -27.42 8.858 0.354
```

The hypothesis is clearly not rejected. Hence, it makes sense to continue the model reduction process, and test hypotheses about the dependence

9.5 Serial correlation and variance homogeneity

of treatment on intercepts and slopes for the regression lines in the usual way.

9.5.4 Model validation

As always it is important to check that the model fits the data appropriately. The major assumptions concern the specification of the fixed effects, variance homogeneity, and the correlation structure.

Residual plots. As usual, residual plots are useful to check the assumptions on the fixed effects as well as the assumption of variance homogeneity. The left part of Figure 9.4 shows a residual plot for mGausNugget, and is easily made with the command

```
> plot(mGausNugget)
```

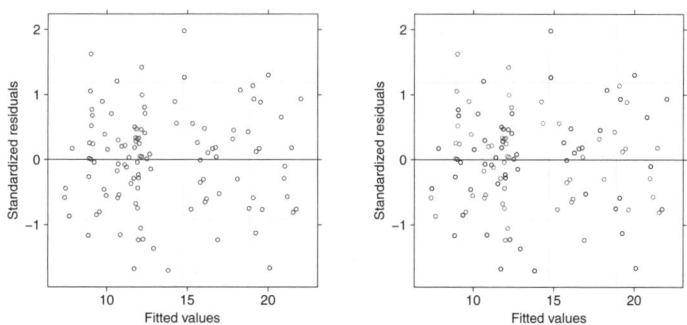

Figure 9.4. Residual plots for the model mGausNugget. The points in the right plot are coloured according to observation day. Residuals from days 26, 45, 61 and 91 are plotted in black, red, green and blue, respectively.

The fitted values on the x-axis includes the estimated fixed effects as well as the estimated (or rather predicted) random effects. In other words, if the model is on the form

$$y_i = \mu_i + u_{g(i)} + e_i$$

with μ_i being the fixed part (here depending on the baseline, treatment and time), $u_{g(i)}$ the random effect, and e_i the error term (containing serial correlation and/or independent noise), then the fitted values are defined as $\hat{y}_i = \hat{\mu}_i + \hat{u}_{g(i)}$. Thus, different values are obtained for different goats. The standardized residuals on the y-axis are defined as

$$\frac{y_i - \hat{y}_i}{\sqrt{\hat{\tau}^2 + \hat{\sigma}^2}}$$

since $\sqrt{\hat{\tau}^2 + \hat{\sigma}^2}$ is the estimated standard deviation of e_i. About 95% of the standardized residuals are thus expected to be between −2 and 2.

Sometimes it is useful to colour the point in the residual plot according to the explanatory variables. In the right plot of Figure 9.4, the points are coloured after observation day, but we could also colour according to feed type or goat:

```
> plot(mGausNugget, col=factor(long.g$time))  # Fig 9.4, right
> plot(mGausNugget, col=long.g$feed)          # Not shown
> plot(mGausNugget, col=long.g$goat)          # Not shown
```

The plot shows a tendency that residuals are larger at day 91 (blue points) compared to the other days. This is not too surprising considering our preliminary investigations of model lmFit.

Notice that fitted values and residuals (unstandardized) can be extracted using mGausNugget$fitted and mGausNugget$residuals. Both objects are matrices with two columns. The second columns are those used by plot above, i.e. $\hat{y}_i = \hat{\mu}_i + \hat{u}_{g(i)}$ and $y_i - \hat{y}_i$. In particular, a plot similar to the left plot in Figure 9.4 could be made with the command

```
> plot(mGausNugget$fitted[,2], mGausNugget$residuals[,2])
```

The first column in mGausNugget$fitted only includes the estimated fixed effects, i.e. $\hat{\mu}_i$, and the first column in mGausNugget$residuals is the corresponding residual, $y_i - \hat{\mu}_i$.

Correlation plot In order to check whether the correlation structure of a given model is appropriate, we may compare the empirical correlations from the lmFit model (see Section 9.4) to those estimated by the model with serial correlation. That is, we will compare the points from Figure 9.2 to the estimated correlation function.

In the left part of Figure 9.5 the comparison is made for mGausNugget. Recall the definition of timeDist and corVec from Section 9.4. Then the plot is made as follows:

9.5 Serial correlation and variance homogeneity

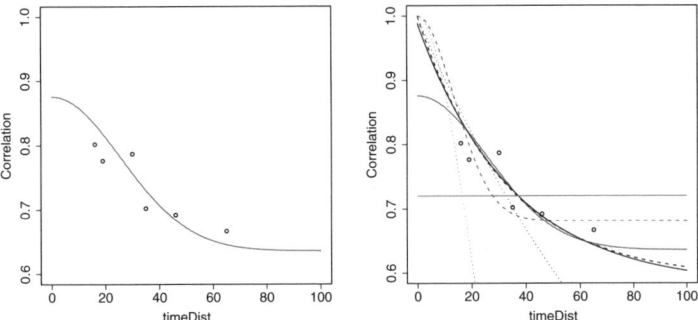

Figure 9.5. Left: Empirical correlations from lmFit (points) together with estimated correlation function for the models mGausNugget (solid red). Right: As the left plot, but with estimated correlation function superimposed for the models mGausNoNugget (dashed red), mGausNoRI (dotted red), mExpNugget (solid blue), mExpNoNugget (dashed blue), mExpNoRI (dotted blue), mRandInt (green).

```
> corFctGausNugget = function(h){
+     (0.1598 + 0.0603 * exp(-(h/36.31)^2))/
+     (0.1598 + 0.0603+0.0313)}    # Estimated corr. function
> plot(timeDist, corVec, xlim=c(0,timeMax),
+      ylim=c(0.6,1), ylab="Correlation")  # The points
> plot(corFctGausNugget, 0, timeMax,
+      add=T, col=2)  # Correlation function in red
```

The curve seems to approximate the points quite well, indicating that the model mGausNugget describes the correlation structure in the data appropriately.

9.5.5 Comparison of different correlation structures

Different models fitted to the same data are often compared with Akaike's Information Criterion (AIC) that takes the model fit as well as the complexity (number of free parameters) of the model into account. The anova command applied to several model computes the AIC values for each of the models (among others):

```
> anova(mGausNugget, mGausNoNugget, mGausNoRI, mExpNugget,
+       mExpNoNugget, mExpNoRI, mRandInt, mUnrestrict)
              Model df    AIC    BIC  logLik  Test L.Rat p-value
mGausNugget       1 21  140.7  194.4  -49.39
mGausNoNugget     2 20  139.8  190.9  -49.93 1 vs 2  1.09  0.2954
```

```
mGausNoRI       3 19 161.3 209.8 -61.65 2 vs 3 23.42  <.0001
mExpNugget      4 21 140.7 194.3 -49.37 3 vs 4 24.54  <.0001
mExpNoNugget    5 20 138.7 189.8 -49.38 4 vs 5  0.00   0.9492
mExpNoRI        6 19 138.8 187.3 -50.41 5 vs 6  2.06   0.1509
mRandInt        7 19 143.6 192.1 -52.81
mUnrestrict     8 25 148.5 212.3 -49.26 7 vs 8  7.10   0.3108
```

The df column gives the number of free parameters in the model, and AIC is computed as -2*logLik+2*df. Models with a small AIC value are to be preferred, so the output suggests that mExpNoNugget is the most appropriate. However, there is unfortunately no theory about the sampling variability in the AIC values, so mExpNoRI and mGausNoNugget may also be good choices for these data. Notice that mRandInt has a quite high AIC value so it is appropriate to incorporate some kind of serial correlation. This is of course as expected. Also, mUnrestrict has a large AIC value so there does not seem to be a need for a very complex structure. The Bayesian Information Criterion (BIC) leads to a similar, although not completely equivalent, conclusion.

Notice that anova output also contain some tests. This makes sense because some of the models are nested. Not all the p-values are reliable, though, so the test results should be used with great care. Unfortunately, the simulate.lme function does not apply to models with a serial correlation structure, so in order to get a more precise approximation to the p-value than the above chi-square approximation, one has to write his or her own code for a simulation study.

We may also want to compare the different correlation structures graphically. This is done in the right part of Figure 9.5. In order to produce this we have defined functions similar to corFctGausNugget above and run the corresponding plot commands. Notice that many of the estimated correlation functions differ very little in the region where there are actually data. Also, notice that one of the models (mGausNoRI, dotted blue) seems to fit the data very badly. Some error must have happened during the numerical optimization in the lme call, and the fit is not trustworthy.

9.6 Models with variance inhomogeneity

In all the models from Section 9.5.1 it was assumed that the variance is the same for all observations. This may be inappropriate, By including a weights option in the call to lme or gls, it is possible to incorporate variance inhomogeneity (heteroskedasticity).

9.6 Models with variance inhomogeneity

Diggle model with time-depending variance. Recall from Section 9.4 that the variance in the goats example seemed to be similar for days 26, 45 and 61, but somewhat larger for day 91. The Diggle model mGausNugget with random effects, Gaussian serial correlation and measurement noise can be extended to allow for variance heteroskedasticity as follows (information about fixed effects has been deleted in the output for clarity):

```
> mDiffVarGaus = lme(weight~w0+factor(feed)*factor(time),
+           random=~1|goat,
+           corr=corGaus(form=~time|goat,nugget=TRUE),
+           weights=varIdent(form=~1|time),
+           data=long.g)

> mDiffVarGaus

[*** Here comes estimates for the fixed effects ***]

Random effects:
 Formula: ~1 | goat
        (Intercept)  Residual
StdDev:   0.3717933  0.256258

Correlation Structure: Gaussian spatial correlation
 Formula: ~time | goat
 Parameter estimate(s):
     range      nugget
 36.4438041  0.4387794
Variance function:
 Structure: Different standard deviations per stratum
 Formula: ~1 | time
 Parameter estimates:
        26         45         61         91
 1.0000000  1.0780677  0.9633404  1.5030406
Number of Observations: 112
Number of Groups: 28
```

With the notation from Section 9.5.4, the variance of the residual term e_i is $\sigma^2 + \tau^2$. It is now allowed to differ between days. The first time point (here 26) is taken as a reference group, so $\hat{\tau}^2 + \hat{\sigma}^2$ is estimated to $0.4388^2 = 0.1925$ at day 26. For day 45, 61 and 91 this estimate should be multiplied by 1.078, 0.9633 and 1.5030, respectively. As expected, the variance is similar at days 26, 45, 61, but larger at day 91. Notice that the nugget effect, i.e. the fraction $\sigma^2/(\tau^2 + \sigma^2)$ is assumed to the same for all time points.

Unrestricted model with time-depending variance. An alternative model with no restrictions on the variance-covariance matrix, i.e., allowing for different variances at different times and freely varying correlations, is fitted with `gls` as follows:

```
> long.g1 = transform(long.g, obsNo=as.numeric(factor(time)))
> head(long.g1)
     goat feed  w0 time weight obsNo
1.26    1    1 20.4   26   21.0     1
1.45    1    1 20.4   45   21.5     2
1.61    1    1 20.4   61   21.3     3
1.91    1    1 20.4   91   22.3     4
2.26    2    1 10.3   26   11.4     1
2.45    2    1 10.3   45   11.6     2

> mDiffVarSymm = gls(weight~w0+factor(feed)*factor(time),
+                    corr=corSymm(form=~obsNo|goat),
+                    weights=varIdent(form=~1|obsNo),
+                    data=long.g1)

> mDiffVarSymm
```

[*** Here comes estimates for the fixed effects ***]

```
Correlation Structure: General
 Formula: ~obsNo | goat
 Parameter estimate(s):
 Correlation:
   1     2     3
 2 0.783
 3 0.705 0.805
 4 0.675 0.699 0.787
Variance function:
 Structure: Different standard deviations per stratum
 Formula: ~1 | obsNo
 Parameter estimates:
       1        2        3        4
1.000000 1.051141 1.017378 1.397968
Degrees of freedom: 112 total; 95 residual
Residual standard error: 0.4290561
```

Notice that `gls` requires the groups defined by the days to be numbered consecutively. Therefore we have made a new dataset (`long.g1`) with a new variable (`obsNo`) that takes the values 1–4 instead of 26, 45, 61, 91. Again, we see that the estimated variances are similar at days 26, 45, and 61, but larger at day 91. The estimated within-goat correlation

9.6 Models with variance inhomogeneity

matrix is reported under 'Correlation Structure'.

As usual we may compare different models with AIC:

```
> anova(mGausNugget, mDiffVarGaus, mDiffVarSymm)
             Model df   AIC   BIC  logLik    Test L.Rat p-value
mGausNugget      1 21 140.7 194.4 -49.39
mDiffVarGaus     2 24 142.6 203.9 -47.32  1 vs 2 4.141  0.2466
mDiffVarSymm     3 27 145.4 214.4 -45.73  2 vs 3 3.161  0.3673
```

The AIC values suggest that there is no need to incorporate variance inhomogeneity for these data.

Unrestricted models with group-depending variance. Variation may differ between groups of subjects rather than between time points as in the following application. The hormone renin (important for regulation of the blood pressure) was measured for 18 liver cirrhosis patients and 6 controls measured under 7 different challenges, here referred to as different measurement times (numbered 1 to 7).[1] The log-concentration of renin is used as response. To get an impression of the variation within the two groups we plot the average profiles in the two groups along with representations of the standard errors (Figure 9.6):

```
> data(renin)
> # Compute means, sd's and number of obs. for each
+ # combination of group and time:
> avedata = with(renin,
+                tapply(l_renin, list(time,grp), mean))
> sddata = with(renin,
+                tapply(l_renin, list(time,grp), sd))
> n_data = with(renin,
+           tapply(l_renin, list(time,grp), length))

> par(mfrow=c(1,2))
> # Points, lower and upper lines in left part
+ # of the plot (patients):
> plot(1:7, avedata[,1], xlim=c(0,8), ylim=c(2,6),
+      xlab="Measurement times", ylab="", type="n")
> lines(1:7, avedata[,1] +
+            1.96*sddata[,1]/sqrt(n_data[,1]))
> lines(1:7, avedata[,1] -
+            1.96*sddata[,1]/sqrt(n_data[,1]))
```

---

[1] The experiment was carried out by Annette Dam Fialla, Department of Medical Gastroenterology, Odense University Hospital.

```
> # Shaded area and curve combining the points in
+ # the left plot
> cord.x = c(1:7,7:1)
> cord.y = c(avedata[,1] -
+          1.96*sddata[,1]/sqrt(n_data[,1]),
+          avedata[7:1,1] +
+          1.96*sddata[7:1,1]/sqrt(n_data[7:1,1]))
> polygon(cord.x, cord.y, col='skyblue')
> lines(1:7,avedata[,1],  type="b", pch=16)

> # Similarly for the right plot (controls):
> plot(1:7,avedata[,2], xlim=c(0,8), ylim=c(2,6),
+     xlab="Measurement times", ylab="", type="n")
> lines(1:7, avedata[,2] +
+       1.96*sddata[,2]/sqrt(n_data[,2]))
> lines(1:7, avedata[,2] -
+       1.96*sddata[,2]/sqrt(n_data[,2]))
> cord.x = c(1:7,7:1)
> cord.y = c(avedata[,2] -
+          1.96*sddata[,2]/sqrt(n_data[,2]),
+          avedata[7:1,2] +
+          1.96*sddata[7:1,2]/sqrt(n_data[7:1,2]))
> polygon(cord.x, cord.y, col='skyblue')
> lines(1:7, avedata[,2], type="b", pch=16)
```

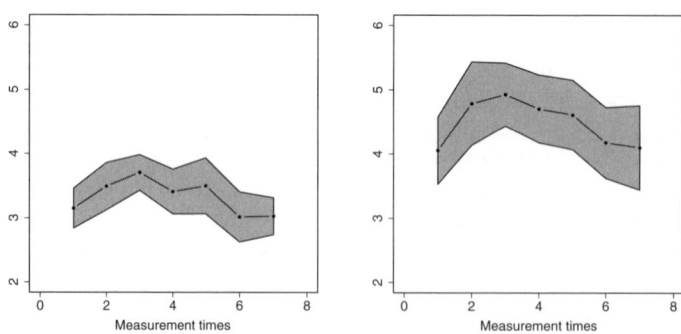

Figure 9.6. Average log-concentration of renin (points) for for the controls (left) and the liver cirrhosis patients (right). The shaded areas represent mean $\pm$ 1.96 times the standard error for each measurement time.

We see that the variation of the hormone (on log-scale) is larger in the patient group than in the control group, and this should be accounted for in the statistical analysis. This can be done in with `gls` as follows:

```
> m1 = gls(l_renin~factor(grp)+factor(time),
```

9.6 Models with variance inhomogeneity

```
+            weights=varIdent(form=~1|grp),
+            correlation=corSymm(form=~time|id), data=renin)
> summary(m1)
Generalized least squares fit by REML
  Model: l_renin ~ factor(grp) + factor(time)
  Data: renin
       AIC      BIC    logLik
  260.5016 355.2452 -99.2508

Correlation Structure: General
 Formula: ~time | id
 Parameter estimate(s):
 Correlation:
   1     2     3     4     5     6
 2 0.759
 3 0.824 0.800
 4 0.866 0.836 0.951
 5 0.853 0.881 0.939 0.975
 6 0.768 0.739 0.853 0.877 0.856
 7 0.702 0.895 0.747 0.805 0.819 0.810
Variance function:
 Structure: Different standard deviations per stratum
 Formula: ~1 | grp
 Parameter estimates:
        1         0
 1.0000000 0.4329904

Coefficients:
                  Value  Std.Error   t-value p-value
(Intercept)    3.0752460 0.20262445 15.177073  0.0000
factor(grp)1   1.1122826 0.31500862  3.530959  0.0005
factor(time)2  0.4827594 0.11646408  4.145135  0.0001
factor(time)3  0.6115495 0.10109264  6.049397  0.0000
factor(time)4  0.3995873 0.08683702  4.601578  0.0000
factor(time)5  0.3574497 0.09159426  3.902533  0.0001
factor(time)6 -0.0416371 0.11441616 -0.363909  0.7164
factor(time)7 -0.1442367 0.13150883 -1.096784  0.2744

Residual standard error: 1.187433
Degrees of freedom: 165 total; 157 residual
```

We use an unstructured correlation structure concerning the 7 measurements from the same subject and allow the variance to be different for the two groups (but the same for each time within the groups). The model can be written as $y_i = \alpha(\text{grp}_i) + \beta(\text{time}_i) + e_i$. The standard deviation of the error term, sd(e_i), is estimated to 1.187 for the patient group and 0.433 for the controls confirming the impression from Figure 9.6 that there is larger variability in the patient group. In the model, two error terms e_i and e_j have correlation zero if they correspond to observations from different subjects. Within subject, the pairwise correlations vary freely, but the correlations are assumed to be the same for both groups. The estimated correlation matrix is listed in the output under 'Correlation structure'.

Assuming the same variance for the two groups is not a good idea as is seen from the below fit (m2), where we obtain an AIC value of 279.6 that should be compared to the one obtained from model m1, where the AIC value is 260.5. This indicates a much better fit of model m1 compared to model m2.

```
> m2 = gls(l_renin~factor(grp)+factor(time),
+          correlation=corSymm(form=~time|id), data=renin)
> summary(m2)
Generalized least squares fit by REML
  Model: l_renin ~ factor(grp) + factor(time)
  Data: renin
       AIC      BIC    logLik
  279.6331 371.3205 -109.8166

[*** Rest of output left out for clarity ***]
```

One can then turn to the interpretation of the estimates of the fixed effects from model m1. Actually it might also be a good idea to investigate whether there is need for interaction between the two factors grp and time, but this will not be pursued any further here.

9.7 Multiple series for each subject

A frequently occurring design when comparing two or more treatments is the within-subject design, comprising cross-over designs. When each treatment results in a series of measurements there are two types of correlation to take into account, namely the correlation within a single series from a single subject and the correlation between measurements from different series from the same subject. While the model for the first is similar to those from section 9.5, we assume here that the correlations between any two measurements from different series are the same. We illustrate this briefly with the data example glucose from the package Guide1data, see Appendix C. Two groups of persons, 12 lean and 12

9.7 Multiple series for each subject

obese, had their glucose response measured at times 0, 30, 60, ..., 360 minutes after an infusion. For each person this was done during three periods with three different types of infusion (A, B or C). Actually, only four persons went through infusion C, and we therefore consider only infusion A and B below. Also we use only the measurements after 210 minutes when a meal was given.[2]

First, after inspection of the data, they are transformed into the long format. Note that the measurement at time 0 is kept (for use as a baseline) while the subsequent measurements are in columns 7 to 18 as indicated in the varying argument of reshape. Also notice the variable carry which is I in period 1, but for period 2 and 3 gives the *previous* infusion type.

```
> library(nlme)
> data(glucose)
> head(glucose)
  subject group infusion period carry   t0  t30  ...  t360
1       1  lean        A      1     I 4.90 4.65  ...  4.73
2       2  lean        A      2     C 4.78 4.71  ...  4.28
3       3  lean        A      2     B 5.10 5.20  ...  5.10
4       4  lean        A      1     I 5.05 5.24  ...  5.29
5       5  lean        A      3     C 4.91 4.86  ...  3.43
6       6  lean        A      2     C 5.24 5.31  ...  5.76
> glucose = subset(glucose, infusion!="C")
> long.gluc = reshape(glucose, idvar="series",
+                     varying=list(7:18), times=30*(1:12),
+                     v.names="gluc", direction="long")
> head(long.gluc)
  subject group infusion period carry  t0 series time gluc
1       1  lean        A      1     I 4.90    1_1   30 4.65
2       2  lean        A      2     C 4.78    2_2   30 4.71
3       3  lean        A      2     B 5.10    3_2   30 5.20
4       4  lean        A      1     I 5.05    4_1   30 5.24
5       5  lean        A      3     C 4.91    5_3   30 4.86
6       6  lean        A      2     C 5.24    6_2   30 5.31
```

To fit the model we need nested random effects of subject and series. That is, on top of the Gaussian correlation structure within series, as specified below, a constant correlation is fitted to measurements from the same subject and another constant is added to correlations from the same series. The within-series correlation structure is the Gaussian model as described in Section 9.5.

---
[2]The data were kindly provided by Birgitte Sloth, Department of Human Nutrition, University of Copenhagen.

```
> mMultiSerGaus = lme(gluc~timefac*infusion*group,
+         random=~1|subject/period,
+         corr=corGaus(form=~time|subject/period,
+         nugget=TRUE),data=long.gluc,
+         subset=time>200)
> summary(mMultiSerGaus)
Linear mixed-effects model fit by REML
 Data: long.gluc
  Subset: time > 200
       AIC      BIC    logLik
  740.1253 843.8279 -341.0627

Random effects:
 Formula: ~1 | subject
        (Intercept)
StdDev:   0.4180308

 Formula: ~1 | period %in% subject
        (Intercept)   Residual
StdDev: 1.240219e-18 0.7667782

Correlation Structure: Gaussian spatial correlation
 Formula: ~time | subject/period
 Parameter estimate(s):
      range        nugget
2.531920e+01 2.582784e-06

Number of Observations: 288
Number of Groups:
         subject period %in% subject
              24                  48
```

The part of the output showing fixed effects estimates is not shown here. We see that in this case the within subject standard deviation (estimated to 1.24e-18) and the measurement error (nugget, estimated to a fraction 2.58e-06) are both negligible. The first of these two shows that we might as well treat different series from the same person as independent, just like series from different persons.

The fixed part of the model could be different in many ways. First, the baseline measure, t0, might be included as covariate. Second, an effect of period might be included, or even a carry-over effect of the previous period, carry. Handling these complicated models is a delicate matter, not only in terms of interpretation, but also in the sense of numerical problems causing the estimation procedure to fail. When successful, however, we may continue by testing the various parts of the fixed effects

9.7 Multiple series for each subject

model, using `anova(reducedModel,fullModel)` as described in Section 9.5.3, but remember to add the option `method="ML"` when fitting the two models to be compared.

10

Generalized linear models

In this chapter we deal with data where the response is binary (0/1), ordinal or a count. Such data should not be analyzed using models based on the normal distribution, such as those dealt with in Chapters 7–9. Still we imagine situations with several explanatory variables so some kind of regression models are needed. Common to the models considered in the present chapter is that they are linear (in the parameters) on some scale, hence the name generalized linear models.

10.1 Logistic regression

We first consider the analysis of a binary response. This means that there are only two possible values of the response variable such as in the following example concerning mortality of ticks in a dose-response trial. The data are available under the name `tickdata` in the package `Guide1data`, see Appendix C.[1] The outcome considered for a given tick is whether or not it is alive after an 11 weeks period. Thus the response variable is

$$Y_i = \begin{cases} 1 & \text{if } i\text{th tick alive} \\ 0 & \text{if } i\text{th tick dead} \end{cases}$$

In the specific example there were two different species of ticks, *orius majusculus* and *orius insidiosus*. Moreover, ticks were treated with different doses of the entomopathogenic fungus *metarhizium anispliae*: Beside zero dose, the following doses (on log10-scale) were applied: 6, 7, 8

---

[1] The experiment was conducted by Susanne Vestergaard, Department of Ecology, University of Copenhagen.

and 9 (spores/ml). Such data can be analyzed using *logistic regression*. The probability of the tick being alive is modeled as a function of species and dose,
$$p = P(Y = 1 | \text{species, dose}),$$
via the logit-link,
$$\text{logit}(p) = \log(p/(1-p)) = \alpha(\text{species, dose}),$$
here allowing for interaction between the two explanatory variables. For the tick data, the ticks were kept in cages of size 10 at the beginning of the experiment and the data are summarized as follows:

```
> data(tickdata)
> head(tickdata)
  experiment    dose surv dead total specie
1          1       0    7    3    10     Oi
2          2       0    6    4    10     Oi
3          3       0    6    4    10     Oi
4          4 1000000    5    5    10     Oi
5          5 1000000    7    3    10     Oi
6          6 1000000    7    3    10     Oi
```

Keeping the ticks in separate cages might give rise to additional clustering which we will return to later but ignore for now. In total we have 129 data lines (not all combinations of dose and species are replicated the same number of times). When data are on "grouped" form as above, then the logistic regression model allowing for interaction is fitted as follows:

```
> fit0 = glm(cbind(surv,dead) ~ factor(specie)*factor(dose),
+            data=tickdata, binomial)
```

Notice how the "response" is given as the number of survived and the numbers of death, cbind(surv,dead). Whether or not the two factors interact can be investigated using the anova function with the test option set to Chisq:

```
> fit1 = glm(cbind(surv,dead) ~ factor(specie)+factor(dose),
+            data=tickdata, binomial)
> anova(fit1, fit0, test="Chisq")
Analysis of Deviance Table

Model 1: cbind(surv, dead) ~ factor(specie) + factor(dose)
Model 2: cbind(surv, dead) ~ factor(specie) * factor(dose)
```

10.1 Logistic regression

```
         Resid. Df Resid. Dev Df Deviance P(>|Chi|)
1            123      322.74
2            119      318.55  4   4.1836    0.3817
> summary(fit1)
Coefficients:
                          Estimate Std.Error z value Pr(>|z|)
(Intercept)                -0.0434   0.1388  -0.313   0.7543
factor(specie)Om            1.2159   0.1332   9.125  < 2e-16 ***
factor(dose)1000000        -0.3682   0.1943  -1.895   0.0581 .
factor(dose)10000000       -1.0381   0.1856  -5.592 2.24e-08 ***
factor(dose)100000000      -1.6549   0.1966  -8.416  < 2e-16 ***
factor(dose)1000000000     -2.8829   0.2500 -11.528  < 2e-16 ***
---
(Dispersion parameter for binomial family taken to be 1)
```

The test for interaction is not significant ($p = 0.38$). There seems to be significant main effects of species and dose, and to get an overall test for dose-effect, for example, one can compare the additive model (fit1) to the model with only main effect of species again using the anova function.

If instead one has the data with one line per tick such as the following

```
> head(ticklong, 15)
   Y dosel speciel experiment
1  1     0      Oi          1
2  1     0      Oi          1
3  1     0      Oi          1
4  1     0      Oi          1
5  1     0      Oi          1
6  1     0      Oi          1
7  1     0      Oi          1
8  0     0      Oi          1
9  0     0      Oi          1
10 0     0      Oi          1
11 1     0      Oi          2
12 1     0      Oi          2
13 1     0      Oi          2
14 1     0      Oi          2
15 1     0      Oi          2
```

with in total 1290 datalines then the above additive model may be fitted by

```
> fit1.long = glm(Y~factor(speciel)+factor(dosel),
+                  data=ticklong, binomial)
> summary(fit1.long)
```

```
Coefficients:
                         Estimate Std.Error z value Pr(>|z|)
(Intercept)               -0.0434    0.1388  -0.313   0.7543
factor(speciel)Om          1.2159    0.1332   9.125  < 2e-16 ***
factor(dosel)1000000      -0.3682    0.1943  -1.895   0.0581 .
factor(dosel)10000000     -1.0381    0.1856  -5.592 2.24e-08 ***
factor(dosel)100000000    -1.6549    0.1966  -8.416  < 2e-16 ***
factor(dosel)1000000000   -2.8829    0.2500 -11.528  < 2e-16 ***
(Dispersion parameter for binomial family taken to be 1)
```

giving us exactly the same parameter estimates (of course). The data are available in this form under the name ticklong in the package Guide1data.

10.1.1 Odds-ratios

Results from a logistic regression model is summarized by odds-ratios. They are easy to get, along with 95%-confidence intervals, for the contrasts reported in the summary(fit1) simply by

```
> exp(coef(fit1))
   (Intercept)       factor(specie)Om    factor(dose)1000000
     0.9574581              3.3733268              0.6919742
factor(dose)10000000    factor(dose)100000000
     0.3540965              0.1911038
factor(dose)1000000000
     0.0559672
> exp(confint(fit1))
Waiting for profiling to be done...
                            2.5 %      97.5 %
(Intercept)              0.72942732 1.25810709
factor(specie)Om         2.60346277 4.39048387
factor(dose)1000000      0.47227130 1.01226795
factor(dose)10000000     0.24534482 0.50819825
factor(dose)100000000    0.12930226 0.27966047
factor(dose)1000000000   0.03367575 0.08998758
```

but you can also choose the ones you would like to calculate using the R package gmodels

```
> library(gmodels)
> cm = rbind('6 vs 0 dose' = c(0,0,1,0,0,0),
+            '9 vs 8 dose'= c(0,0,0,0,-1,1),
+            'Oi vs Om'= c(0,-1,0,0,0,0))
```

10.1 Logistic regression

```
> exp(estimable(fit1, cm, conf.int=0.95))[,c(1,6,7)]
             Estimate Lower.CI  Upper.CI
6 vs 0 dose 0.6919742 0.4705877 1.0175115
9 vs 8 dose 0.2928628 0.1766881 0.4854241
Oi vs Om    0.2964433 0.2275809 0.3861422
```

From the last output we see that odds of surviving the considered time period is 70% smaller for *orius insidiosus* as compared to *orius majusculus* (odds-ratio is estimated to 0.296) in a situation with the same level of dose for the two species. The odds of surviving the considered time period is also 70% smaller for ticks (of same species) given log10-dose equal to 9 as compared to log10-dose equal to 8. There is also a reduced odds of surviving when comparing log10-dose 6 to no dose, but it is not significant at the 5% significance level as the 95%-confidence interval contains the value 1.

10.1.2 Correlated data

Due to the experimental conditions in the above example with ten ticks kept in the same cages, there may be additional clustering (or variation) in the data than what is accounted for by dose and species. This is also sometimes referred to as overdispersion and may be accounted for as follows:

```
> fit1.cl = glm(cbind(surv,dead)~factor(specie)+factor(dose),
+              data=tickdata, quasibinomial)
> summary(fit1.cl)
Coefficients:
                        Estimate Std. Error t value Pr(>|t|)
(Intercept)              -0.0434     0.2133  -0.204 0.838849
factor(specie)Om          1.2159     0.2046   5.941 2.70e-08 ***
factor(dose)1000000      -0.3682     0.2984  -1.234 0.219718
factor(dose)10000000     -1.0381     0.2851  -3.641 0.000398 ***
factor(dose)100000000    -1.6549     0.3020  -5.480 2.30e-07 ***
factor(dose)1000000000   -2.8829     0.3841  -7.506 1.07e-11 ***
(Disp. parameter for quasibinomial family taken to be 2.359)
```

As is seen we get the same estimates but the estimated standard errors are larger now acknowledging the additional variation in the data. Therefore *p*-values have also become larger, but with no major effect on the overall conclusions about the effect of the two factors on survival. The overdispersion parameter is estimated to 2.359 and the corrected standard errors are simply obtained as the uncorrected ones multiplied by $\sqrt{2.359}$. It is also possible to calculate the odds-ratios taking the overdispersion into account,

```
> exp(estimable(fit1.cl, cm,conf.int=0.95))[,c(1,6,7)]
             Estimate  Lower.CI  Upper.CI
6 vs. 0 dose 0.6919742 0.3832567 1.2493673
9 vs 8 dose  0.2928628 0.1350104 0.6352742
0i vs 0m     0.2964433 0.1977028 0.4444986
```

and we see that 95%-confidence intervals have become wider.

Another way of taking the correlation into account is by use of the approach of generalized estimating equations (GEE). Often one will have data on long form (corresponding to one line per tick) so we here show the GEE analysis using the ticklong data frame. To use the gee function we need the gee package, and the rows of the data need to be ordered with respect to the clustering variable (experiment). This is already the case the tick data, but we show here how to make the required ordering anyway:

```
> data(ticklong)
> ord = order(as.integer(ticklong$experiment))
> ticklong = ticklong[ord,]
> fit.gee = gee(Y~factor(speciel)+factor(dosel),
+               data=ticklong, id=experiment,
+               family=binomial, corstr="independence")
> summary(fit.gee)
GEE:  GENERALIZED LINEAR MODELS FOR DEPENDENT DATA

Coefficients:
                              Est NaiveSE RobustSE Robust z
(Intercept)               -0.04347  0.1396   0.2057  -0.2112
factor(speciel)Om          1.21589  0.1339   0.2068   5.8781
factor(dosel)1000000      -0.36820  0.1954   0.3222  -1.1425
factor(dosel)10000000     -1.03818  0.1866   0.2689  -3.8600
factor(dosel)100000000    -1.65493  0.1977   0.2955  -5.5989
factor(dosel)1000000000   -2.88298  0.2514   0.3243  -8.8894
```

The point estimates are still unchanged, but now two sets of estimated standard errors are reported. The naive ones correspond to those reported when running the logistic regression model fit1 (although they are not exactly the same) and the robust ones are those taking the correlation into account so they are the one to be used — note they are close to those reported when using the overdispersion correction, fit1.cl.

10.1 Logistic regression

Using a cluster corrected analysis will not always result in larger estimated standard errors. In a cross-over study, taking the clustering into account will typically result in a smaller standard error for the estimated treatment effect.

10.1.3 Natural response: calculating it yourself!

In another experiment some other ticks (of the same species) were exposed to different doses of an entomopathogenic fungus, and survival after some period of time was of interest.[2] The data are summarized in the table:

	Ticks exposed to an entomopathogenic fungus	
Dose	Number of ticks	Number dead
0	60	40
10^6	30	22
10^7	60	51
10^8	60	56
10^9	60	59

We will now try to model what is sometimes referred to as *natural response*, that is, that some of the ticks will die in the considered time period even though they are given zero dose. Let p_j denote the probability that a tick from dose group j will die in the given time period. The natural response model is formulated as

$$p_j = \begin{cases} c & j = 1 \text{ (dose zero)} \\ c + (1-c)\tilde{p}_j & j > 1 \end{cases}$$

where

$$\tilde{p}_j = \frac{\exp[\alpha + \beta \cdot \log 10(d_j)]}{1 + \exp[\alpha + \beta \cdot \log 10(d_j)]}$$

with d_j being the dose given in dose group j. Let Y_j denote the number of dead ticks and n_j the number of ticks in dose group j, $j = 1, \ldots, 5$. Assume that it is reasonable to use the binomial model to describe these data, hence, $Y_j \sim \text{binom}(n_j, p_j)$. This model with natural response is

---

[2] This experiment was also carried out by Susanne Vestergaard, Department of Ecology, University of Copenhagen.

not easily analyzed with any standard function in R such as for example glm. However, there is a built-in function in R to do maximum likelihood estimation. All we need is the log-likelihood function which in this case is given by

$$l(c, \alpha, \beta) = \sum_{j=1}^{5} \log \left(\binom{n_j}{Y_j} \right) + Y_j \log (p_j) + (n_j - Y_j) \log (1 - p_j)$$

In R we specify the (minus) log-likelihood function as follows:

```
> dose = c(0,10^6,10^7,10^8,10^9)  # The data
> numb = c(60,30,60,60,60)
> numbd = c(40,22,51,56,59)

# And now minus log-likelihood function:
> logl = function(c,al,be){
+   res=log(choose(numb[1],numbd[1]))+
+   numbd[1]*log(c)+(numb[1]-numbd[1])*log(1-c)
+   for (i in 2:5){
+     p=c+(1-c)*exp(al+be*log10(dose[i]))/
+     (1+exp(al+be*log10(dose[i])))
+     res=res+
+       log(choose(numb[i],numbd[i]))+
+     numbd[i]*log(p)+(numb[i]-numbd[i])*log(1-p)
+   }
+   return(-res)
+ }
```

There are at least two functions in R to do maximum likelihood estimation, we choose the one in the R package bbmle called mle2:

```
> library(bbmle)
> guess = list(c=0.2, al=0, be=0)  # Starting values
> opt = mle2(logl, start=guess, method="Nelder-Mead")
> summary(opt)
Maximum likelihood estimation
Coefficients:
     Estimate Std. Error z value   Pr(z)
c    0.665638  0.059189  11.2460 < 2.2e-16 ***
al  -9.384604  3.681482  -2.5491  0.010799 *
be   1.359110  0.470933   2.8860  0.003902 **
---
-2 log L: 17.20422
```

Notice that −2logL is given so it possible to compare this model to the model where dose is used as a factor:

10.2 Proportional odds model

```
> fit.dose = glm(cbind(numbd,numb-numbd) ~ factor(dose),
+                binomial)
> -2*logLik(fit.dose)[1]
[1] 17.13965
>   17.20422 + 2*logLik(fit.dose)[1]   # Test statistic
[1] 0.06456612
> 1-pchisq(0.065, df=2)                # p-value
[1] 0.9680224
```

Hence the model with natural response seems to give a fine fit to the data. From summary(opt) we see that there is a significant dose effect since β (be) is significantly different from zero.

10.2 Proportional odds model

In the following dataset the interest is on studying some treatments (restrictions on feeding) effect on chickens leg problems. Leg problems are summarized with a so-called gait score on an ordered scale from 0 to 4 with 0 corresponding to no leg problems while 4 corresponds to serious problems. Such a response variable is called ordinal. The chickens were exposed to one of the following four treatments: food ad libitum (1), fasting 8 hours each day in a week (2), fasting 8 hours each day in two weeks (3), and fasting 8 hours each day in three weeks (4). The data used for this example are available as gaitdata in the package Guide1data, see Appendix C.

An ordinal response may be analyzed by grouping the response variable into only two categories, say (0,1,2) and (3,4), and proceeding with a logistic regression. This approach does not take fully account of the information in the data, however, and one often applies the so-called proportional odds model instead. This model is in the present data setting formulated as follows. Let

$$\gamma_j(\text{trt}_i) = P(Y_i \leq j | \text{trt}_i), \quad j = 0, 1, 2, 3,$$

where Y_i denotes the ordinal score, trt_i the treatment for the ith chicken, and j gives the categories of the response score. It is then assumed that the odds are given by

$$\frac{\gamma_j(\text{trt}_i)}{1 - \gamma_j(\text{trt}_i)} = \exp\left(\theta_j + \alpha(\text{trt}_i)\right)$$

so that

$$\text{logit}(P(Y_i \leq j | \text{trt}_i)) = \theta_j + \alpha(\text{trt}_i), \quad j = 0, 1, 2, 3.$$

The assumption that the α's do not depend on j gives rise to *proportional odds*: the odds-ratio for two chickens that received treatment 2 and 1, say, becomes

$$OR_{2:1} = \exp\left(\theta_j + \alpha(2) - \theta_j - \alpha(1)\right) = \exp\left(\alpha(2) - \alpha(1)\right),$$

so the odds are proportional and independent of the considered category j. This is an assumption which can and should be tested in practice; we return to this below.

There are at least two functions in R, `polr` and `vglm`, that can fit this model, but as far as we know only the `vglm` function can be used to test the proportional odds assumption. To fit the proportional odds model using the `vglm` function we first need to load the VGAM package:

```
> data(gaitdata)
> library(VGAM)
> pomod = vglm(gait~factor(treat),
+              family=cumulative(parallel=T), data=gaitdata)
[1] "head(extra$orig.w)"
NULL
> summary(pomod)
Coefficients:
                 Value Std. Error  t value
(Intercept):1  -1.4379849   0.23824 -6.035990
(Intercept):2   0.0596624   0.22174  0.269062
(Intercept):3   1.6420000   0.25226  6.509232
(Intercept):4   4.6763496   0.73061  6.400588
factor(treat)2 -0.0046428   0.29978 -0.015487
factor(treat)3  0.6487115   0.30167  2.150366
factor(treat)4  0.8618298   0.30478  2.827693

Log-likelihood: -394.5315 on 1177 degrees of freedom
```

Before interpreting the above parameter estimates, we check whether or not the proportional odds assumption is reasonable in the present case:

```
> multmo = vglm(gait~factor(treat),
+               family=cumulative(parallel=F), data=gaitdata)
[1] "head(extra$orig.w)"
NULL
> test.po = -2 * (logLik(pomod)-logLik(multmo)) # Test stat.
> test.po
[1] 10.33626
> df.po = length(coef(multmo)) - length(coef(pomod)) # Df
> 1-pchisq(test.po, df=df.po)                        # p-value
[1] 0.3239547
```

Notice that the option `parallel=T` has been replaced by `parallel=F` in order to fit the model without proportional odds. The proportional odds assumption seems to be appropriate ($p = 0.32$).

We can thus use the estimates from `pomod` to calculate relevant estimated odds ratios, for example, the odds-ratio $\widehat{OR}_{4:1}$ comparing treatment 4 to treatment 1 (the control):

```
> exp(0.862)
[1] 2.367892
> exp(0.862-1.96*0.305) #Lower 95% confidence limit
[1] 1.302389
> exp(0.862+1.96*0.305) #Upper 95% confidence limit
[1] 4.305098
```

We see that the odds of having no leg problems is more than twice as big (2.36) for a chicken receiving treatment 4 compared to a chicken from the control group. By the way, this is also the odds ratio of having gait-score below or equal to 1, or below or equal to 2, see the definition of $\gamma_j(\text{trt}_i)$. If we wish an overall test of the treatment effect, we can compare the proportional odds model fit to the fit of the model leaving out the treatment factor:

```
> po.int = vglm(gait~1, family=cumulative(parallel=T),
+               data=gaitdata)
[1] "head(extra$orig.w)"
NULL
> test.treat = -2 * (logLik(po.int)-logLik(pomod))
> test.treat
[1] 13.26311
> 1-pchisq(test.treat, df=3)
[1] 0.004100827
```

We see that there treatment effect is clearly significant.

As a last note on the proportional odds model one may use the function `repolr` in the case of an ordinal response with clustered data such as repeated recordings from the same units.

10.3 Poisson regression

Data where the response variable consists of counts may be analyzed using Poisson regression.

As illustration we use the following data about bees.[3] At a certain date (July 1st, 2010) some traps (actually plates) intended for catching bees were checked, and the number of bees on each plate and the type of bee, solitary bee or bumblebee, were registered. The plates were of different colors (blue, white and yellow) and were placed at four different locations in Denmark (Havreholm, Kragevig, Saltrup and Svaerdborg). In total there were 36 plates. The data, beedata in the package Guide1data, see Appendix C, look like

```
> beedata
    Locality Replicate  Color  Time       Type Number id
1  Havreholm         A  White july1 Bumblebees      1  1
2  Havreholm         B  White july1 Bumblebees      2  2
3  Havreholm         C  White july1 Bumblebees      1  3
4  Havreholm         A Yellow july1 Bumblebees      2  1
5  Havreholm         B Yellow july1 Bumblebees      2  2
6  Havreholm         C Yellow july1 Bumblebees      0  3
.
. (more data lines here)
.
70  Kragevig         A   Blue july1   Solitary      3  4
71  Kragevig         B   Blue july1   Solitary      1  5
72  Kragevig         C   Blue july1   Solitary      4  6
```

It was of special interest to see whether the two types of bees "preferred" plates of different colors. We therefore fit a model with main effect of locality and with interaction between type of bee and color of plate:

```
> fit1 = glm(Number~
+           factor(Locality)+factor(Color)*factor(Type),
+           data=beedata, family=poisson)
> summary(fit1)
Coefficients:
                               stimate   SE      z p-value
(Intercept)                      -1.06 0.71  -1.49 0.136
factor(Locality)Kragevig         -0.47 0.24  -1.97 0.048 *
factor(Locality)Saltrup          -1.86 0.40  -4.58 0.000 ***
factor(Locality)Svaerdborg       -1.86 0.40  -4.58 0.000 ***
factor(Color)White                1.38 0.79   1.75 0.079 .
factor(Color)Yellow               1.87 0.75   2.46 0.013 *
factor(Type)Solitary              2.56 0.73   3.49 0.000 ***
factor(Color)Wh:factor(Type)Sol  -1.75 0.84  -2.06 0.038 *
factor(Color)Yel:factor(Type)Sol -2.13 0.81  -2.61 0.008 **
```

---

[3]The data was collected by Casper Ingerslev Henriksen, Department of Agriculture and Ecology, University of Copenhagen.

10.3 Poisson regression

Before going into interpretation of parameter estimates we may wish to check whether the included interaction term is significant. This can be investigated using the anova function with the Chisq option:

```
> fit2 = glm(Number~
+           factor(Locality)+factor(Color)+factor(Type),
+           data=beedata, family=poisson)
> anova(fit2, fit1, test="Chisq")
Analysis of Deviance Table

Model 1: Number ~ factor(Locality) +
         factor(Color) + factor(Type)
Model 2: Number ~ factor(Locality) +
         factor(Color) * factor(Type)
  Resid. Df Resid. Dev Df Deviance P(>|Chi|)
1        65     93.956
2        63     84.215  2   9.7403  0.007672 **
```

We see that the interaction term is indeed significant, meaning that the two species have different colour preferences.

To summarize results from a Poisson regression analysis one reports relative risks, that is, ratios of estimated means. This can be done in a similar way as we did for the logistic regression analysis using the R package gmodels:

```
> library(gmodels)
> cm = rbind(
+   'Havreholm vs.Kragevig'= c(0, -1, 0, 0, 0, 0,0,0,0),
+   'Havreholm vs.Saltrup'= c(0, 0, -1, 0, 0, 0,0,0,0),
+   'Saltrup vs. Svaerdborg'= c(0, 0, 1, -1, 0, 0,0,0,0),
+   'Soli vs bumblebies,col=blue'= c(0, 0, 0, 0, 0, 0,1,0,0),
+   'Soli vs bumblebies,col=yellow'= c(0, 0, 0,0, 0, 0,1,0,1),
+   'Soli vs bumblebies,col=white'= c(0, 0, 0,0, 0, 0,1,1,0),
+   'Yellow vs blue, bumblebies'= c(0, 0, 0, 0, 0, 1,0,0,0),
+   'White vs blue, bumblebies'= c(0, 0, 0, 0, 1, 0,0,0,0),
+   'Yellow vs blue, soli'= c(0, 0, 0, 0, 0, 1,0,0,1),
+   'White vs blue, soli'= c(0, 0, 0, 0, 1, 0,0,1,0))

> exp(estimable(fit1, cm, conf.int=0.95))[,c(1,6,7)]
                                 Estimate  Lower.CI   Upper.CI
Havreholm vs.Kragevig           1.6071429 0.9969207   2.590886
Havreholm vs.Saltrup            6.4285714 2.8710147  14.394399
Saltrup vs. Svaerdborg          1.0000000 0.3462905   2.887749
Soli vs bumblebies, col=blue   13.0000000 3.0316698  55.744857
Soli vs bumblebies, col=yellow  1.5384615 0.7587886   3.119266
```

Soli vs bumblebies, col=white	2.2500000	0.9684140	5.227619
Yellow vs blue, bumblebies	6.5000000	1.4403251	29.333655
White vs blue, bumblebies	4.0000000	0.8334583	19.197122
Yellow vs blue, soli	0.7692308	0.4263705	1.387798
White vs blue, soli	0.6923077	0.3767949	1.272018

We see that bumblebees are more often caught in yellow plates compared to blue plates with an estimated relative risk equal to 6.5 with corresponding 95%-confidence interval (1.44,29.33). This preference for yellow plates is absent for the solitary bees where the corresponding relative risk and 95%-confidence interval are 0.77 and (0.43,1.39), respectively.

In the experiment the plates were actually placed in clusters of size three (with each of the three colors represented). There were three clusters (replicates A, B and C) at each location giving us in total 12 clusters. The cluster id's are given in the variable id. It is possible to take this structure into account either by using the option family=quasipoisson in the glm-call (this is the overdispersion approach similar to what we did in the logistic regression case) or by using the GEE approach. Let us illustrate the latter one here. Again, to use the gee function, we need the gee package, and the rows of the data must be ordered with respect to the clustering variable, in this case the id-variable. It is also possible to calculate the relative risks similar to what we did above.

```
> ord = order(as.integer(beedata$id))
> beedata = beedata[ord,]
> fit.gee = gee(Number~
+               factor(Locality)+factor(Color)*factor(Type),
+               data=beedata, id=id, family=poisson,
+               corstr="independence")
> exp(estimable(fit.gee, cm, conf.int=0.95))[,c(1,6,7)]
```
	Estimate	Lower.CI	Upper.CI
Havreholm vs.Kragevig	1.6071429	0.8007259	3.225708
Havreholm vs.Saltrup	6.4285714	2.7253787	15.163592
Saltrup vs. Svaerdborg	1.0000000	0.4206828	2.377088
Soli vs bumblebies, col=blue	13.0000000	3.0075778	56.191398
Soli vs bumblebies, col=yellow	1.5384615	0.6464261	3.661461
Soli vs bumblebies, col=white	2.2500000	1.1121004	4.552197
Yellow vs blue, bumblebies	6.5000000	1.4515260	29.107298
White vs blue, bumblebies	4.0000000	1.0283565	15.558806
Yellow vs blue, soli	0.7692308	0.4392505	1.347104
White vs blue, soli	0.6923077	0.3373783	1.420631

The estimates are unchanged but the 95%-confidence intervals changed

10.3 Poisson regression

(a little) now taking the cluster structure into account.

11
Non-linear regression

In this chapter we show how to fit non-linear regression models to data where the response is continuous. We refer to Ritz and Streibig [2008] for more details on how to do non-linear regression analysis using R.

11.1 Non-linear regression with `nls`

We illustrate non-linear regression analysis using the data `orgmat` in `Guide1data` concerning the influence of two antibiotics, Fenbendazole and Levamisole, on the decomposition of dung organic matter from cattle.[1] Dung was collected two days after treatment and mesh bags with

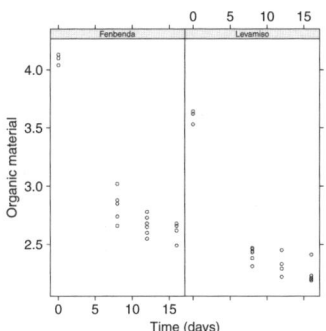

Figure 11.1. Organic material plotted against time (days).

---

[1]The experiment was conducted by Christian Sommer, Department of Ecology, University of Copenhagen as part of a larger study.

portions of 40 g of dung were placed in the soil. The organic matter was then determined after 0, 8, 12 and 16 weeks. A plot of the data, see Figure 11.1, reveals that the contents of organic matter seems to be non-linear as a function of time.

```
> data(orgmat)
> library(lattice)
> xyplot(mat~time|antibio, xlab="Time (days)",
+        ylab="Organic material'',type="p", data=orgmat)
```

Let us first consider the data where the antibiotic Fenbendazole was applied. An appropriate relationship of the amount of organic matter and time t might be the function

$$OM(t) = a + be^{-ct}$$

with a, b and c unknown parameters. This is a non-linear function of the parameters. The statistical model where we add normally distributed errors can be fitted using the non-linear least squares function nls. To have a successful fit one often needs to supply starting values for the parameters to nls as the estimation procedure is iterative and may diverge without sound starting values. One way to proceed is to "guess" starting values using the behavior of the non-linear function as illustrated in the following. Since $OM(0) = a + b$ and OM converges to a for t tending to infinity this leads to an idea of the parameters a and b. Then we can evaluate the function at a specific value of t, e.g. $OM(8) \approx 2.75$, to get an idea of the last parameter c. Hence we obtain

$$a \approx 2.6, \ b \approx 4.1 - 2.6 = 1.5, \ c \approx -\frac{1}{8}\log\left(\frac{2.75 - 2.6}{1.5}\right) = 0.29,$$

and can then fit the model:

```
> fit.Fen = nls(mat~a+b*exp(-c*time),
+               start=list(a=2.6,b=1.5,c=0.29),
+               data=subset(orgmat,antibio=="Fenbenda"))
> summary(fit.Fen)
Formula: mat ~ a + b * exp(-c * time)

Parameters:
  Estimate Std. Error t value Pr(>|t|)
a  2.56049    0.05750   44.53  < 2e-16 ***
b  1.52982    0.07563   20.23 2.49e-13 ***
c  0.21713    0.03356    6.47 5.77e-06 ***
Residual standard error: 0.09026 on 17 degrees of freedom
```

11.1 Non-linear regression with nls

Estimates, standard errors and p-values are read as usual.

There exists some built-in self-starter functions, and, in fact, the function we used above is available under the re-parameterization

$$OM(t) = a^* + (b^* - a^*)e^{-e^{c^*}t}$$

with $a^* = a$, $b^* = a+b$, $c^* = -\log(c)$. This self-starter function is called SSasymp, and the model can then be fitted *without giving starting values* as follows:

```
> fit.tmp = nls(mat~SSasymp(time,a.st,b.st,c.st),
+               data=subset(orgmat,antibio=="Fenbenda"))
> summary(fit.tmp)
Formula: mat ~ SSasymp(time, a.st, b.st, c.st)
Parameters:
     Estimate Std. Error t value Pr(>|t|)
a.st   2.5605    0.0575  44.531  < 2e-16 ***
b.st   4.0903    0.0521  78.508  < 2e-16 ***
c.st  -1.5272    0.1546  -9.881 1.85e-08 ***
Residual standard error: 0.09026 on 17 degrees of freedom
```

We now also include data from the second antibiotic and fit a model where the parameters a, b, and c may be different for the two antibiotics. This model is a sub-model of the two-way ANOVA model with interaction between times used a grouping variable (factor) and antibiotics type. We can thus start by making a lack-of-fit test:

```
> fit0 = lm(mat~antibio*factor(time), data=orgmat)
> fit.nls =
+     nls(mat~a[antibio]+b[antibio]*exp(-c[antibio]*time),
+         data=orgmat,
+         start=list(a=c(2.57,2.21), b=c(1.5,1.4), c=c(0.23,0.23)))
> anova(fit.nls, fit0)
Analysis of Variance Table

Model 1: mat ~ a[antibio] + b[antibio] *
                exp(-c[antibio] * time)
Model 2: mat ~ antibio * factor(time)
  Res.Df Res.Sum Sq Df   Sum Sq F value Pr(>F)
1     33    0.22756
2     31    0.22522  2 0.0023404  0.1611  0.852
```

Note how we let the parameters depend on the antibiotics in the non-linear regression analysis. Hence a, b and c are vectors of length 2. It is further seen that the lack-of-fit test is clearly non-significant. We can then test the hypothesis that, for example, the two c-coefficients are identical as follows:

```
> fit.nls1 =
+       nls(mat~a[antibio]+b[antibio]*exp(-c*time),
+           data=orgmat,
+           start=list(a=c(2.57,2.21), b=c(1.5,1.4), c=0.23))
> anova(fit.nls1, fit.nls)
Analysis of Variance Table

Model 1: mat ~ a[antibio] + b[antibio] * exp(-c * time)
Model 2: mat ~ a[antibio] + b[antibio] *
                 exp(-c[antibio] * time)
  Res.Df Res.Sum Sq Df   Sum Sq F value Pr(>F)
1     34    0.23006
2     33    0.22756  1 0.002496   0.362 0.5515
```

This seems to be plausible. Similarly for the other sets of parameters:

```
> fit.nls21 = nls(
+       mat~a[antibio]+b*exp(-c*time), data=orgmat,
+       start=list(a=c(2.21,2.57), b=c(1.44), c=0.23))
> fit.nls22 = nls(
+       mat~a+b[antibio]*exp(-c*time), data=orgmat,
+       start=list(a=c(2.35), b=c(1.38,1.51), c=0.23))
> anova(fit.nls1, fit.nls21)
Analysis of Variance Table

Model 1: mat ~ a[antibio] + b[antibio] * exp(-c * time)
Model 2: mat ~ a[antibio] + b * exp(-c * time)
  Res.Df Res.Sum Sq Df    Sum Sq F value Pr(>F)
1     34    0.23006
2     35    0.24923 -1 -0.019173  2.8336 0.1015
> anova(fit.nls1, fit.nls22)
Analysis of Variance Table

Model 1: mat ~ a[antibio] + b[antibio] * exp(-c * time)
Model 2: mat ~ a + b[antibio] * exp(-c * time)
  Res.Df Res.Sum Sq Df    Sum Sq F value    Pr(>F)
1     34    0.23006
2     35    0.88381 -1 -0.65376  96.619 1.809e-11 ***
```

The conclusion is that there is a clear need for two separate a-coefficients (the asymptotes) while there seems to be no need for the two other

11.2 Model validation, transform-both-sides

parameters (b and c) to be different for the two antibiotics. A plot of the fitted values, see Figure 11.2, can be obtained as follows:

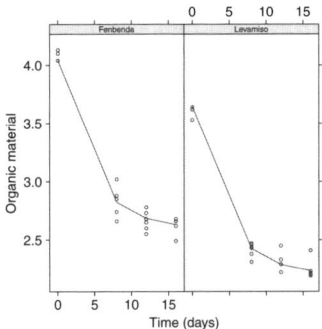

Figure 11.2. Organic material plotted against time (days) together with predicted values (combined with lines).

```
> co = coef(fit.nls21)
> xyplot(mat~time|antibio, xlab="Time (days)", type="p",
+        data=orgmat)
> trellis.focus("panel",1,1) # Access row 1 and column 1
> with(subset(orgmat, antibio=="Fenbenda"),
+        panel.lines(time, co[1]+co[3]*exp(-co[4]*time)))
NULL
> trellis.focus("panel",2,1) # Access to column 2, row 1
> with(subset(orgmat, antibio=="Levamiso"),
+        panel.lines(time, co[2]+co[3]*exp(-co[4]*time)))
NULL
> trellis.unfocus()
```

Note how the function trellis.focus allows us to plot in the separate plotting windows created by xyplot.

11.2 Model validation, transform-both-sides

In the current example with a few different times and replicates at each each time point, we could do model validation based on the two-way ANOVA model in two steps: First, the two-way ANOVA is fitted and validated in the usual way (see Section 7.3). Second, the non-linear regression model is tested against the two-way ANOVA as we did above. With many different values of the regressor and without replicates this is not an option and in such a situation we need to validate the non-linear regression model directly.

For illustration, let us consider the data from the Levamisole group only. Unfortunately, standardized residuals cannot be calculated based on an nls object so we have to make a plot of the ordinary (or raw) residuals,

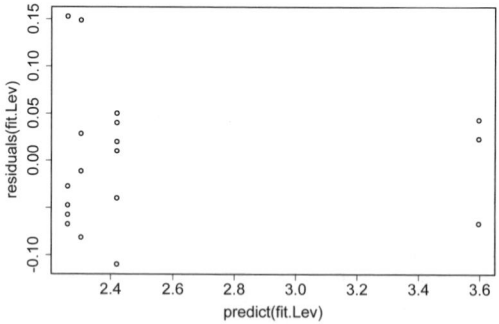

Figure 11.3. Residual plot for the Levamisole group from the data on decomposition of organic material.

see Figure 11.3:

```
> fit.Lev = nls(mat~a+b*exp(-c*time),
+               start=list(a=2.6,b=1.5,c=0.29),
+               data=subset(orgmat,antibio=="Levamiso"))
> plot(predict(fit.Lev), residuals(fit.Lev))
```

The residual plot shows some indication of variance heterogeneity with seemingly larger variation for small predicted values. A solution to this, as we have seen previously, might be to transform the response variable. However, for a non-linear regression where the non-linear function is often based on substance knowledge, a transformation of the response, only, will distort this relationship and the interpretation of the parameters. It is therefore more appealing to transform both sides of the non-linear equation. This is referred to as the transform-both-sides approach. For example, if we for some reason wanted to use the square root transformation, we would modify the nls command to

```
> nls(sqrt(mat)~sqrt(a+b*exp(-c*time)),
+     start=list(a=2.6,b=1.5,c=0.29),
+     data=subset(orgmat,antibio=="Levamiso"))
```

A Box-Cox variant (power transformations, see also Section 7.3.2) of this

11.2 Model validation, transform-both-sides

approach is available in the package **nlrwr** as shown below.

```
> library(nlrwr)
> fit.Lev1 = boxcox.nls(fit.Lev)  # TBS Box-Cox plot
> bcSummary(fit.Lev1)

Estimated lambda: 1.9
Confidence interval for lambda: [-0.73,    NA]
> summary(fit.Lev1)

Formula: bcFct1(mat) ~ bcFct2(a + b * exp(-c * time))
Parameters:
  Estimate Std. Error t value Pr(>|t|)
a  2.23353    0.04432   50.39  < 2e-16 ***
b  1.36327    0.05325   25.60 2.06e-14 ***
c  0.24877    0.03845    6.47 7.74e-06 ***
Residual standard error: 0.1672 on 16 degrees of freedom
```

The boxcox.nls call produces Figure 11.4. From the output we see that the confidence interval for the power transformation parameter contains 1 so there is not sufficient evidence for using a power transformation, and we can base our conclusions on the original (untransformed) analysis. You may compare the above summary(fit.Lev1) to summary(fit.Lev), which not shown here, to see that there is virtually no difference between the two analyses in this case.

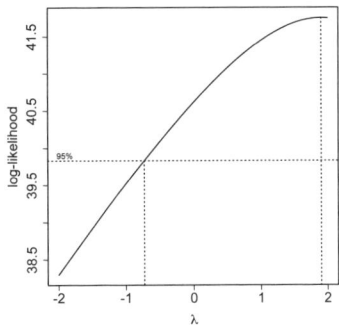

Figure 11.4. Log-likelihood function for the power parameter in the transform-both-sides Box-Cox approach, based on data from the Levamisole group.

11.3 Estimation of derived parameters

Sometimes it may be of interest to estimate a function of the parameters in a given model. As an illustration say it is of interest to estimate the derivative of the function $OM(t)$ at time $t = 16$ and also to give a 95%-confidence interval. Since the derivative of OM is $OM'(t) = -cbe^{-ct}$, we are interested in

$$OM'(16) = -cbe^{-c16}.$$

Notice that this is a non-linear function of the parameters b and c, which were found not to be significantly different for the two antibiotics. We use the estimates from `fit.nls21`. To estimate the above non-linear function and compute its associated standard error one uses the so-called *delta method*, which is implemented in the package `alr3`. It is used as follows:

```
> library(alr3)
> delta.method(fit.nls21, "-c*b*exp(-c*16)")
                   Estimate           SE
-c*b*exp(-c*16) -0.008345624 0.002621577
> -0.00835 + 1.96*0.00262  #Lower limit in 95% conf. interval
[1] -0.0032148
> -0.00835 - 1.96*0.00262  #Upper limit in 95% conf. interval
[1] -0.0134852
```

Notice how the estimates and standard errors in the two last commands are used to compute the 95% confidence intervals, giving us $(-0.0135, -0.0032)$.

12
Survival analysis

In this chapter we consider the analysis of survival data, that is, data where the response is the time until some event occurs, often the death of a subject. For such data it is common that not all responses are observed — for example, some subjects may still be alive at end of study, this is called right-censoring. Special statistical techniques are then required.

12.1 Survival data

Survival analysis or failure time data analysis means the statistical analysis of data, where the response of interest is the time, T, from a well-defined time origin to the occurrence of some given event (end-point). The key example is the time from randomization to a given treatment until death occurs, leading to the observation of survival times. In behavioral studies in agricultural science, one may for example observe the time from a domestic animal has received some stimulus until it responds with a given type of action. Standard methods will often be inappropriate because survival times are frequently incompletely observed with the most common example being right-censoring. This may be because the patient is still alive at the point in time where the study is closed and the data are to be analyzed, or because the subject is lost for follow-up due to other reasons.

As an example we use the PBC data available in the `survival` package. It concerns the survival of 418 liver cirrhosis patients. There are many explanatory variables recorded for these patients, but we will here focus only on the effect of edema and serum albumin. The variable `edema` is either 0 (no edema), 0.5 (untreated or successfully treated), or 1 (edema despite diuretic therapy). The variable `albumin` is the serum albumin concentration (mg/dl). We will make use of functions available in the package `timereg` so we start by loading that package.

```
> library(survival)
> library(timereg)
> data(pbc)
> pbc1 = with(pbc,data.frame(id,time,status,edema,albumin))
> head(pbc1)
  id time status edema albumin
1  1  400      2   1.0    2.60
2  2 4500      0   0.0    4.14
3  3 1012      2   0.5    3.48
4  4 1925      2   0.5    2.54
5  5 1504      1   0.0    3.53
6  6 2503      2   0.0    3.98
```

The variable `time` gives the number of days between registration and the earlier of death, transplantation, or end of study in July 1986, and `status` is the status at endpoint (0/1/2 for censored/transplant/dead). We will here also take transplant as censored, corresponding to collapsing groups 0 and 1. Notice, however, that a more complete analysis would be to study competing risks as this is really the setting in the current situation.

12.2 Kaplan-Meier estimator, log-rank test

As mentioned above, survival data are often incomplete, typically in form of a right-censored version of the survival times such as in the

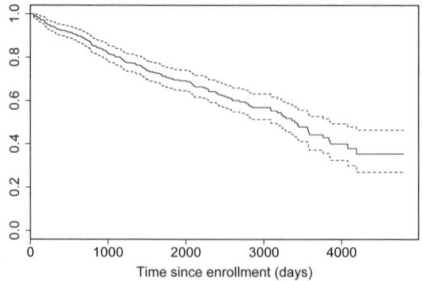

Figure 12.1. Kaplan-Meier plot for the complete PBC dataset.

12.2 Kaplan-Meier estimator, log-rank test

PBC example. It turns out that even though one does not fully observe the survival times, one can still estimate the distribution of the survival times. This is typically done in terms of estimating the survival function,

$$S(t) = P(T > t),$$

and the aim is to examine how the survival function depends on the explanatory variables (here edema and albumin). The survival function is estimated by the Kaplan-Meier estimator. Thinking of the PBC data as a single sample (not taking edema and albumin into account), the Kaplan-Meier estimator is constructed and plotted as follows:

```
> fit.all = survfit(Surv(time,status==2) ~ 1, data=pbc1)
> plot(fit.all,
        xlab="Time since enrollment (days)", mark.time=F)
```

Notice how the response (to the left of the "tilde") is written. The result is shown in Figure 12.1 along with 95% (pointwise) confidence limits. Notice that the estimator jumps at the time points where a death has taken place. The Kaplan-Meier estimator can also be calculated for different populations as those defined by the variable **edema**:

```
> fit.edema = survfit(Surv(time,status==2)~edema, data=pbc1)
> plot(fit.edema, xlab="Time since enrollment (days)",
+       ylim=c(0,1), col=c("red","green","blue"), mark.time=F)
> legend(3500,0.9,legend=c("edema 0","edema 1/2","edema 1"),
+       text.col=c("red","green","blue"))
```

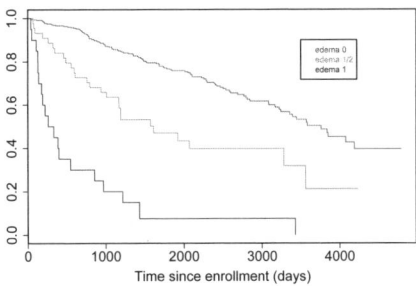

Figure 12.2. Kaplan-Meier plot for the PBC data stratified after edema.

The plot is shown in Figure 12.2. We see that edema seems to be a serious risk indicator since the estimated survivor function for the group with **edema** equal to 1 is consistently lower compared to the two other groups. The probability of surviving 1000 days for a patient with edema present (**edema** equal to 1) is estimated to roughly 0.20, while for a patient without edema (**edema** equal to 0) it is estimated to around 0.90.

The log-rank test is a test that can compare the survival of these three groups. It is calculated as follows

```
> survdiff(Surv(time,status==2) ~ edema, data=pbc1)
Call:
survdiff(formula = Surv(time, status == 2) ~ edema,
    data = pbc1)

             N Observed Expected (O-E)^2/E (O-E)^2/V
edema=0    354      116   145.47      5.97      62.3
edema=0.5   44       26    13.05     12.84      14.0
edema=1     20       19     2.47    110.44     113.1

Chisq= 131  on 2 degrees of freedom, p= 0
```

The null hypothesis of no difference between the survival of these three groups is clearly rejected as the p-value is very small (zero).

12.3 The Cox proportional hazards model

It is usually desirable to quantify differences in survival among groups. The method of choice for this is the Cox proportional hazards model. Based on the proportional hazards assumption one can report relative risks summarizing the effect of a given explanatory variable.

Model fit. The Cox model is fitted as follows, here using only **edema** as the only explanatory variable:

```
> cox.edema = coxph(Surv(time,status==2) ~ factor(edema),
+                   data=pbc1)
> summary(cox.edema)
  n= 418
                  coef exp(coef) se(coef)      z Pr(>|z|)
factor(edema)0.5 0.9317     2.538   0.2175  4.283 1.84e-05 ***
factor(edema)1   2.3325    10.303   0.2543  9.174  < 2e-16 ***
```

The effect of edema is still rendered significant, and we can see that the relative risk (comparing a patient with edema to one without, i.e.

12.3 The Cox proportional hazards model

level 1 to level 0) is estimated to be $e^{2.33} = 10.3$. We can also include other potentially important variables such as albumin (here included on log-scale):

```
> cox.all = coxph(Surv(time,status==2) ~
+                 factor(edema)+log(albumin), data=pbc1)
> summary(cox.all)
  n= 418
                  coef exp(coef) se(coef)      z Pr(>|z|)
factor(edema)0.5 0.703     2.020   0.2201  3.196  0.00139 **
factor(edema)1   1.737     5.684   0.2767  6.279 3.40e-10 ***
log(albumin)    -3.826     0.021   0.5886 -6.500 8.02e-11 ***
```

Now the relative risk (comparing a patient with edema to one without, and both with the same value of albumin) is estimated to be 5.7 still being significantly different from 1 (the logarithmic relative risk reported in the summary is significantly different from 0). The estimated relative risk associated with `log(albumin)` corresponds to the increased risk by increasing log(albumin) with 1 while keeping edema fixed. This effect is also highly significant, and since the estimate is negative, albumin is protective.

Validating the proportional hazards assumption. The nice summary in terms of estimated relative risks from the Cox analysis is based on the proportional hazards assumption. This assumption should be validated, and many procedures have been advocated for this purpose. Below is shown the Cox score process plot which can be obtained via the `cox.aalen` function from the package `timereg`:

```
> fit.cox = cox.aalen(Surv(time,status==2)~
+             prop(factor(edema))+prop(log(albumin)),
+             data=pbc1)
> summary(fit.cox)
Test for Proportionality
                       sup| hat U(t) |  p-value H_0
prop(factor(edema))0.5            5.26         0.144
prop(factor(edema))1              5.02         0.048
prop(log(albumin))                1.69         0.260

> par(mfrow=c(2,2))
> plot(fit.cox, score=T, xlab="Time (days)")
```

Figure 12.3 displays the observed score process plots for edema (top) and log-albumin (bottom). The bold curves are the observed score processes,

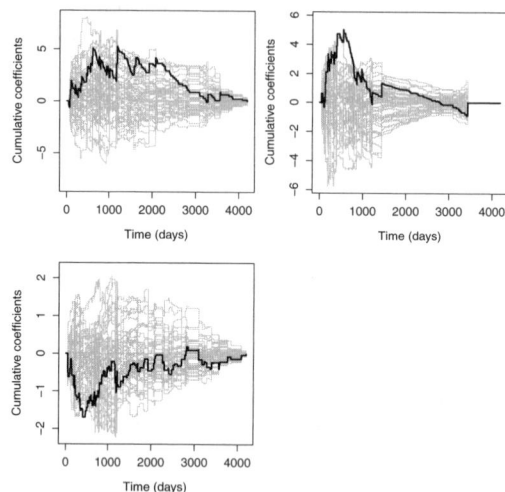

Figure 12.3. Score process plots for the PBC data. The plots in the top are for edema level 0.5 and 1, respectively. The plot in the bottom is for log-albumin.

whereas the grey curves are 50 resampled processes under the null that the proportional hazards assumption holds. A deviating observed process indicates that the assumption may not hold. This might be the case for the upper right display. A specific test of the null can be obtained using the summary function which indeed suggests a problem concerning the edema-group, $(p = 0.048)$. A typical violation of the assumption is that the effect of the variable changes with time. This is explored further in the next section.

12.4 Time-varying effects

An alternative to the Cox model is the Aalen additive hazards model. An appealing feature of this model is that it readily allows for regression effects to change with time, that is, regression coefficients are now taken to be functions of time. It is easy to estimate the cumulated regression coefficients and these are the ones reported by the function aalen in R. They are estimated and plotted as follows:

```
> fit.aalen = aalen(Surv(time,status==2) ~
+       factor(edema)+log(albumin), data=pbc1, max.time=3000)
> summary(fit.aalen)
Additive Aalen Model
```

12.4 Time-varying effects

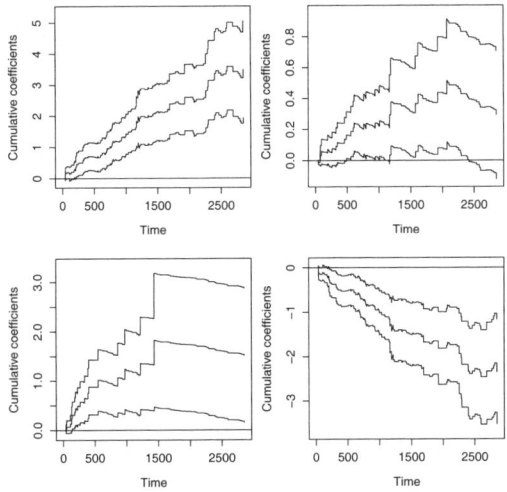

Figure 12.4. Cumulative regression coefficients in the Aalen additive hazards model fitted to the PBC data for the intercept (top left), edema level 0.5 (top right), edema level 1 (bottom left), and log-albumin (bottom right).

```
Test for nonparametric terms

Test for non-significant effects
                Sup-test of significance p-value H_0: B(t)=0
(Intercept)                         5.34                0.000
factor(edema)0.5                    2.67                0.083
factor(edema)1                      3.35                0.009
log(albumin)                        4.92                0.000

Test for time invariant effects
                Kolm-Smir test p-value H_0:const. effect
(Intercept)                0.508                  0.722
factor(edema)0.5           0.302                  0.012
factor(edema)1             1.060                  0.001
log(albumin)               0.441                  0.574

> par(mfrow=c(2,2))
> plot(fit.aalen)
```

Figure 12.4 displays the estimated cumulative regression coefficients and 95% pointwise confidence limits. Since a straight line corresponds to a constant effect, we see that the effect of albumin (on log-scale) seems to be constant with time whereas the effect of edema (level 0.5 and 1) seems to level off after approximately 700 days. This is confirmed by the reported test for constant effect giving us the p-values 0.012 and 0.001, respectively. The above statistical approach is a so-called nonparametric method, and it gets unstable after approximately 3000 days because only few patients are left at risk. Hence we stop the analysis there using the max.time option.

An alternative, somewhat simpler approach is to use the Cox model but so that the effect of edema is allowed to change after some pre-specified timepoint(s), here we will take $t = 700$ for illustration. This can obtained using the survSplit function as shown in the following.

```
> pbc2 = survSplit(pbc1, cut=c(700), end="time",
+                  start="start", event="status")
> pbc2$edemanew = pbc2$edema * as.numeric(pbc2$time>700)
> cox.all.new = coxph(Surv(start,time,status==2) ~
+      factor(edema)+log(albumin)+factor(edemanew),
+      data=pbc2)
> summary(cox.all.new)
                    coef exp(coef) se(coef)     z Pr(>|z|)
factor(edema)0.5   1.464     4.326    0.362  4.036 5.43e-05 ***
factor(edema)1     2.381    10.818    0.368  6.456 1.07e-10 ***
log(albumin)      -3.813     0.022    0.596 -6.389 1.67e-10 ***
factor(edemanew)0.5 -1.117   0.327    0.462 -2.419   0.0156 *
factor(edemanew)1   -1.300   0.272    0.582 -2.233   0.0255 *
```

The two last estimates, -1.12 and -1.30, give the *change* of the effects of edema level 0.5 and edema level 1, respectively. Both are seen to be significant. The effect of edema1 after 700 days is given by $2.38 - 1.30 = 1.08$. This estimate and its standard error can be obtained as shown below:

```
> library(alr3)
> coefVec = as.vector(coef(cox.all.new))
> vcovMat = cox.all.new$var
> names(coefVec) = LETTERS[1:5]
> delta.method(coefVec, "B+E",vcovMat)
    Estimate        SE
B+E 1.080598 0.4734302
> exp(c(1.081-1.96*0.473, 1.081, 1.081+1.96*0.473))
[1] 1.166398 2.947626 7.449002
```

The latter line gives the estimated relative risk, 2.95, and the corresponding 95%-confidence interval. It is clearly seen that the effect of edema has diminished after 700 days.

A
How to install R

To install R under Microsoft Windows do as follows:

- Go to http://cran.r-project.org
- Click **Download R for Windows**
- Click **base**
- Click **Download R 2.13.1 for Windows** (or whatever the newest version is) and save the file, for example on the desktop.
- When the file has been downloaded, then double-click it, accept the license, and install the program.

For Mac users, click **Download R for MacOS X**, then **R-2.13.1.pkg** (version number may change), and follow the instructions.

For Linux users, click **Download R for Linux**, choose your Linux distribution, and follow the instructions.

If you are asked to choose a CRAN Mirror, then choose Denmark.

B
Add-on packages

The base package of R contains most of the functions that we need, such as `lm`, `anova`, `summary`, etc. In these notes we have, however, also used functions from add-on packages. There is a large number of such R packages available which are not automatically installed with the base package. A package is basically a collection of R functions. For example, we have used `lme` from the `nlme` package and `estimable` from the `gmodels` package.

To access the functions from a package, the package should once and for all be installed on your local computer. On a computer with internet connection, click "Packages" in the R menu and you get the option "Install package(s) from CRAN" which lists the possible packages. Click the wanted package and it is installed! This only needs to be carried out once on your computer. To actually use the functions from the add-on package (and for the help information to be visible) the package must also be loaded, either via the "Packages" menu or by the `library` command

```
> library(packagename)
```

where `packagename` is the name of the R package. You need to do this every time R is re-started.

Some packages, for example `gmodels` and `nlme`, are already installed with the base package, but still need to be loaded in each R session. Hence, all you need to do for these to packages, is to write

```
> library(gmodels)
> library(nlme)
```

Just to give you an idea of the development of R, we mention that there are currently more than 3000 packages available at CRAN!

C
The data package Guide1data

The datasets used in the second part of this book are gathered in an R package which is called Guide1data. This makes it easy for you to try the examples yourself without having to type the data. However, you need to install the package, which is in the file Guide1data_1.0.zip. To install it you

- copy the file to your computer,
- Click the Package menu in R, choose "Install from local zip-file", and then browse to find the package file.

Now the package is installed on your computer, but recall that each time you start R you need to *load the package*, if you are going to use the data from the package. This may be done from the menu or by the command library(Guide1data).

In order to use a particular dataset, say organic, write

```
> data(organic)
```

To get information on the content of the dataset, write

```
> ?organic
```

The package currently contains the datasets named

```
bees, chlorophyll, chocolate, endometriosis, goats,
hydrolysis, milkyield, niacin, organic, orgmat, porkers,
renin, streptococcus, vitE.
```

D

Getting help and further information

To get help on how to use a certain function in R, for example the function `read.table`, you may simply write

```
> ?read.table
```

and R will open a window with some text about how to use the function. However, this only works when you know what function to use and remember its name. Quite often you remember that something can be done, but you have forgotten how. The reference card by Short [2004] is very well suited for this purpose. If, for example, you have forgotten the name of the normal distribution function, you find it under the heading "Distributions" in the reference card. There is also, of course, a chance that you may find it in the present guide. You can also use two question marks for a broader search, for example,

```
> ??normal
```

and you will find the normal distribution under the entry `stats::Normal` meaning that you find it in the `stats` package, which is automatically loaded. Hence, you can continue writing

```
> ?Normal
```

and you will see the appropriate help page.

Notice also that in general other useful help pages may be found from each help page through the "See also" links.

To expand your knowledge in R it is usually better to use either a book on R, or one of the numerous R guides or manuals on the web. Thus, until you are experienced you may find *An Introduction to* R [W.N. Venables, 2011] useful. To find it, go to the R project home page, http://www.r-project.org/, and click "Manuals". You can also reach the R project home page from the help menu in R. You may either download the pdf version or browse the HTML version. For example, if you are looking for the normal distribution you may click on "Probability distributions" and you will jump to the right section.

Books specifically on doing statistical analyses using R are Venables and Ripley [2002], Crawley [2007], Dalgaard [2002], Everitt and Hothorn [2010], and Fox and Weisberg [2011]. If you need material on the use of R as general purpose programming language, the book by Venables and Ripley [2000] is recommendable.

Finally, remember to share your knowledge with other R users.

Bibliography

M.J. Crawley. *The R book*. Springer, 2007.

P. Dalgaard. *Introductory Statistics with R*. Springer, 2002.

B.S. Everitt and T. Hothorn. *A Handbook of Statistical Analyses Using R*. Chapman & Hall/CRC Press, 2010.

J. Fox and S. Weisberg. *An R Companion to Applied Regression*. SAGE Publications, 2011.

R Development Core Team. *R: A Language and Environment for Statistical Computing*. R Foundation for Statistical Computing, Vienna, Austria, 2010. URL http://www.R-project.org/. ISBN 3-900051-07-0.

C. Ritz and J.C. Streibig. *Nonlinear Regression with R*. Springer, 2008.

T. Short. *R Reference Card*, 2004. URL cran.r-project.org/doc/contrib/Short-refcard.pdf.

W.N. Venables and B.D. Ripley. *S programming*. Springer, 2000.

W.N. Venables and B.D. Ripley. *Modern Applied Statistics with S*. Springer, fourth edition, 2002.

D.M. Smith W.N. Venables. *An Introduction to R*, 2011. URL cran.r-project.org/doc/manuals/R-intro.pdf. Notes on R: A Programming Environment for Data Analysis and Graphics Version 2.13.1 (2011-07-08).

Index

R console window, 11
R script, 12
drop1, 75
ks.test, 51
NA, 29
SSasymp, 147
Surv, 155
VGAM, 138
VarCorr, 94
aalen, 159
anova, 70, 71, 95
bcSummary, 151
boxcox.nls, 151
boxcox, 85
cex.axis, 45
cex.lab, 45
cex, 45
chisq.test, 57
col, 45
confint, 72
corr, 110
coxph, 156
c, 24
delta.method, 152
dim, 21
estimable, 88
fisher.test, 60
gee, 134
getVarCov, 113

glm, 130
gls, 112
gmodels, 88
head, 21
identify, 47
interaction.plot, 77
intervals, 95
is.na, 29
kruskal.test, 65
lattice, 105
legend, 46
length, 41
lines, 45
lme4, 98
lmer, 99
lme, 94, 106
lm, 70
locator, 47
main, 45
matrix, 29
max, 42
mean, 42
median, 42
min, 42
mle2, 136
names, 21
nlme, 94, 106
nlrwr, 151
nls, 147

nugget, 112
pairs, 109
pch, 45
plot, 45
pnorm, 50
points, 45, 46
power.prop.test, 67
power.t.test, 65
predict, 83
pwr.anova.test, 66
qnorm, 50
qqline, 51
qqnorm, 51, 83
read.xlsx, 20
relevel, 72, 86
rep, 25
reshape, 104
residuals, 83
rnorm, 50
rstandard, 83
savePlot, 47
sd, 42
seq, 25
shapiro.test, 51
simulate.lme, 96
subset, 33
sum, 29, 42
survSplit, 160
survdiff, 156
survfit, 155
t.test, 51
tapply, 105
transform, 37
trellis.focus, 149
var, 42
vglm, 138
which.max, 42
which.min, 42
with, 37
write.xlsx, 20
xlab, 45
xlsx, xlsReadWrite, 20

xyplot, 105

Aalen additive hazards model, 158
adjusted means, 88
AIC, 117
anova, 69
    one-way anova, 69
    two-way anova, 73
argument, 17
average profiles, 105

baseline, 106
binomial distribution, 54
Box-Cox analysis, 84, 150

chi-square test
    simple hypothesis in the multi-
        nomial distribution, 57
    test of homogeneity, 57
correlation matrix, 108
correlation plot, 116
correlation structure, 107
Cox proportional hazards model, 156

data frame, 21
default argument value, 18
delta method, 152
Diggle model, 110, 119

exponential serial correlation, 111

Fisher's exact test, 59

Gaussian serial structure, 111
GEE, 134, 142
generalized linear models, 129

help system, 167
heteroskedasticity, 118

installation of R, 161
interaction, 73

Kaplan-Meier estimator, 155
Kappa statistic, 61

INDEX

Kolmogorov-Smirnov test, 51
Kruskal-Wallis test, 64

linear normal models, 69
linear regression, 78
log-rank test, 156
logical values, 27
logistic regression, 129
long form, 104
longitudinal data, 103

matrix, 29
    determinant, 30
    multiplication, 30
McNemar's test, 60
missing values, 28
model validation, 83, 93, 115, 149
muitiple linear regression, 79
multinomial distribution, 55

non-linear regression, 145
normal distribution, 49
normal quantile plot, 50

odds ratio, 132
overdispersion, 133

plot, 45
Poisson distribution, 55
Poisson regression, 139
power calculation, 65
profile plots, 104
proportional odds model, 137

QQ-plot, 51
quadratic regression, 79

random effects, 91
    nested random effects, 97
    non-nested random effects, 98
random intercepts model, 106
repeated measurements, 103
residual plot, 83, 93, 115
residuals

linear model, 83
mixed model, 93
repeated measures, 116

sample size calculation, 65
score process plot, 157
self-starter function, 147
serial correlation, 110
Shapiro-Wilks tests, 51
subject profile, 104
summary measure, 105
survival analysis, 153

t-test
    paired, 51
    unpaired, 52
test of homogeneity, 57
test of normality
    Kolmogorov-Smirnov test, 51
    Shapiro-Wilks tests, 51
transform-both-sides, 150

unrestricted correlation, 111

variable, 16
variance inhomogeneity, 118
variance-covariance matrix, 108, 113

wide form, 104
Wilcoxon rank sum test, 62
Wilcoxon signed rank test, 63
within-subject design, cross-over design, 124